P9-DNC-830

THE MICROBE FILES

(*with answers*)

Cases in Microbiology for the Undergraduate

Marjorie Kelly Cowan
Miami University of Ohio

Benjamin
Cummings

San Francisco   Boston   New York
Capetown   Hong Kong   London   Madrid   Mexico City
Montreal   Munich   Paris   Singapore   Sydney   Tokyo   Toronto

Publisher: Daryl Fox

Sponsoring Editor: Amy Folsom Teeling

Associate Project Editor: Erin Joyce

Managing Editor: Wendy Earl

Production Supervisor: Sharon Montooth

Cover and Interior Design: tani hasegawa

Copyeditor: Anita Wagner

Compositor: The Left Coast Group, Inc.

Manufacturing Supervisor: Stacey Weinberger

Marketing Manager: Lauren Harp

Library of Congress Cataloging-in-Publication Data

Cowan, M. Kelly.
    The microbe files : cases in microbiology for the undergraduate with answers / M. Kelly Cowan.
        p. cm.
    Includes index.
    ISBN 0-8053-4927-8
        I. Medical microbiology--Case studies. I. Title.

QR46.C867 2001
616'.01--dc21

2001017163

ISBN: 0-8053-4927-8

10 9 8 7 6 5 4 3 -MAL-05 04 03 02

Dedication

*This book is dedicated with love and gratitude
to my sons and my husband.*

About the Author

 Marjorie Kelly Cowan, Ph.D., teaches
microbiology and epidemiology at the
Middletown campus of Miami University in
Ohio. Her training took place at the
University of Louisville, the University of
Groningen (the Netherlands), and the
University of Maryland. She actively studies
teaching and learning, and oversees a
research program in microbial adhesion and
colonization in medical and industrial set-
tings. She recently received a Celebration of Teaching Award from
the Greater Cincinnati Consortium of Colleges and Universities.
She has two sons, Taylor, 12, and Sam, 9.

Acknowledgments

Thank you to my students, who continuously teach me how to teach and show me why it is worth doing. Many of the cases in this book originated in stories students told me about their own experiences in health care.

Of course many people contributed in direct and indirect ways to this project. The book came about through the vision of Amy Folsom Teeling (Sponsoring Editor) and the hard work of Erin Joyce (Associate Project Editor) at Benjamin Cummings. Jeff Howard came up with the title. Sharon Montooth shepherded the book through production. Thanks also to Jim Lewellyn and Suzanne Evans at Miami University for proofreading and assistance. I want to publicly acknowledge the leadership and support of Anne Morris Hooke, chair of microbiology at Miami University. Her humanity and dedication to excellence, fairness, and integrity, are an inspiration to all who know her.

This book was much improved by the comments and suggestions of a multitude of reviewers. Their passion for teaching was evident in their thoughtful reading and commentary, and I have deep appreciation for their input: Cindy Anderson, Mt. San Antonio Community College; Rick Bliss, Yuba College; Burton Clark, Oregon Institute of Technology; Diane Dudzinski, Washington State Community College; Christine Frazier, Southeast Missouri State; Alan Gillen, Pensacola Christian College; Hank Harris, Iowa State University; Leslie Lichtenstein, Massasoit Community College; Katie Morrison-Graham, Lane Community College; Carmen Rexach-Zellhoefer, Merced College; Leba Sarkis, Aims Community College; Phillip Taylor, Glenville State College; Karin VanMeter, Des Moines Area Community College; Ruth Wrightsman, Saddleback College; and Susan Wyckoff, Bradley University.

Finally, heartfelt thanks to my husband Paul Wehner. What a world it would be if everyone had just one person like Paul to believe in them.

Marjorie Kelly Cowan

Contents

Introduction

Dear Student,

When you are learning about infectious diseases, the amount of information you have to master may seem overwhelming. There are so many infections, so many microorganisms, so much new terminology. This book is designed to help your brain organize those long lists of facts into logical pathways. You'll be solving cases instead of rehashing the same old facts. Once you've placed a piece of information into the right place in the big picture, it will have a meaning for you that will be almost impossible to forget!

The Microbe Files

The six chapters in this book provide examples of infectious diseases presented to you the way you will encounter them in your life or clinical practice. The cases are grouped by chapter according to the body site that is most affected. This mirrors what happens in clinical practice: your first encounter with a patient's infection will be what you can see and what the patient can tell you.

In the News Cases

You will find an "In the News" case in every chapter. If you've picked up a newspaper since you've started studying microbiology you may have noticed that nearly every day you can find an article about a new infection, a new epidemic, or a new treatment for a disease. And now you can read these with a knowing eye! The "In the News" cases in this book will get you started in applying your expertise.

Challenge Cases

Each chapter also contains at least one "Challenge" case. Like the other cases in the book, these should require no further information than that obtained in your course. They will ask you to push yourself and to apply what you know within a new context or with a different twist.

The Usual Suspects

Each chapter contains a box that lists the most common infectious agents causing symptoms in the body system under consideration. Don't refer to the list right away; but if you feel stumped, flip to it to see if there's something you missed in your mental browsing.

Glossary

You will also find a glossary in the back of the book that contains many terms that are either very important or particularly mysterious. It is easy to forget the glossary is there, so to remind you to use it, each glossary word appears in **boldface** type the first time it appears in a case.

Steps in Diagnosing a Patient

Anatomical Diagnosis

Clinical diagnosis has a pattern, and typically it goes something like this: a patient comes to you with obvious signs or symptoms, or has a verbal explanation of invisible body sensations. (This is called the *presenting* complaint or symptom.) He or she wants a diagnosis. Even though we have enormous amounts of diagnostic technology available to us, it is obviously impractical (not to mention unethical!) to give patients every available diagnostic test every time they experience illness of any type. The inevitable conclusion is that reaching a diagnosis requires complex critical skills on the part of the practitioner (you) to determine which, if any, diagnostic tests must be performed. Right away, the presenting symptoms allow you to narrow the possibilities. The symptoms will most likely affect a particular area of the body most obviously—like the skin, or the digestive tract. At this point your list of hundreds of possible infections becomes a list of a handful of possible infections. It should be noted that one whole category of diseases seems to affect the entire body. These infections are called **systemic,** and once again, knowing that will help you narrow your choices. Identifying the physical site of symptoms is called the **anatomical diagnosis** and is the first step in clinical diagnosis.

Differential Diagnosis

At this point you have a short list of possible diseases—infectious and noninfectious—that may be causing the symptoms. The list of diseases that could be causing the anatomical symptoms is called the **differential diagnosis.** The differential diagnosis may contain only two or three diseases. If the presenting symptom in a 5-year-old patient is a sore throat with no other symptoms, *Streptococcus pyogenes* or viruses (as a group) are probably the only two causes in the differential diagnosis. On the other hand, a single differential diagnosis may list several infectious agents plus a few metabolic disorders. Consider the symptom of jaundice: it could be caused by one of several hepatitis viruses (which must be sorted out) as well as alcoholism or inherited disorders. On the other end of the spectrum, in a few infections the outward signs are so unique to a particular microorganism that only one infectious cause is possible. The bull's-eye rash of Lyme disease, when present, is diagnostic of *Borrelia burgdorferii* infection, for example.

Etiological Diagnosis

Once you've got the list narrowed to a differential diagnosis, it is time to make the **etiological diagnosis.** In this phase, the actual causative organism is identified, if it is an infectious disease. In the case of jaundice, blood would be drawn and immunologic tests conducted for each of the hepatitis viruses. Other causes can be investigated with blood samples as well (by examining levels of liver enzymes, etc). Sometimes culture of the infected site must be performed.

Epidemiology

Midway through the process of generating the differential diagnosis, and continuing into the etiological diagnosis phase, **epidemiological considerations** become indispensable. This means that you need to ask the following questions related to diseases you are going to list in the differential diagnosis: What people are at risk for this disease? Do particular behaviors (traveling, smoking, etc.) expose one to the disease? (Has this patient engaged in these behaviors?) What is the geographical distribution of the disease? (Has the patient been in these locales?) Have I seen a lot of this disease recently in my practice? Has the patient been immunized against this disease? (Is there a possibility for vaccine failure?) Of course, the answers to many of these questions are obtained from the patients themselves, and clinicians have to be discerning about the patient's ability to "self-report." So in addition to being an epidemiologist you must be a bit of a psychologist as well.

Diagnosis in Practice

In practice, the first two phases—anatomical and differential diagnosis—occur very quickly. The third phase often takes time and money, which is reason enough to make good guesses in the earlier phases. It is important to note that in practice the third phase is often not necessary at all. This occurs for several reasons. If all of the possible etiological agents would be treated in the same way, or not at all (viral rhinitis, for instance), there is no need to identify the organism. Also, the decision not to identify the organism can be determined by a delicate balance between the need to know and the degree of discomfort or expense to the patient. A very common example of this last situation is the pediatric ear infection (otitis media). Typically the physician prescribes treatment solely on the basis of clinical signs, even though several different bacteria can cause the infection. Obtaining samples from the middle ear for culture is unpleasant for the patient, and in the past physicians found that simply prescribing a broad-spectrum antibiotic was the best approach for handling these infections. Now that there is great concern about antibiotic resistance, physicians often opt for "watchful waiting" instead of prescribing an antibiotic in the case of ear infections. But even in this scenario, an etiological diagnosis is usually not made. Ear infections of all types usually resolve themselves within a few days.

For some common conditions, written flow diagrams exist for making all of these decisions and treating the patient ("If this symptom is present, perform the following tests"). These are sometimes called algorithms, or clinical pathways, and may be written by hospital administration and be part of hospital policy. Insurance companies also have diagnosis and treatment algorithms. But it is difficult to predict all possible scenarios, and most of the time diagnosis still requires your own good judgment and experience. Accordingly, some of the cases in this book illustrate what you, as a student of the health sciences and of biology, already know. Health, because it is a biological phenomenon, has a lot of gray areas. There is no single rule of diagnosis. The pattern of diagnosis described here is just that—a pattern. You will diverge from that pattern when your clinical judgment tells you it is appropriate. The cases in this book are designed to help you learn the pattern, and learn how to determine when to veer from it.

Practice Flying

I tell my students that exams cause anxiety because they ask you to fly solo—to answer questions with a minimum of background information available to you. Therefore, I tell them, you have to *practice flying* while you study. Pose problems to yourself without giving yourself access to the answers. Don't give up at the first twinge of difficulty and flip to the answer in your notes, or in the back of the book. Use this opportunity while you're still on the ground to push yourself further. Experience the discomfort of racking your brain now, and later, when it counts, you will be less likely to have to. It will pay off on the exam, and more importantly, it will pay off in your practice. You can think of this book as your flight simulator!

Best of luck,
Marjorie Kelly Cowan

THE MICROBE FILES

with answers

Diseases of the skin and Eyes

Chapter 1

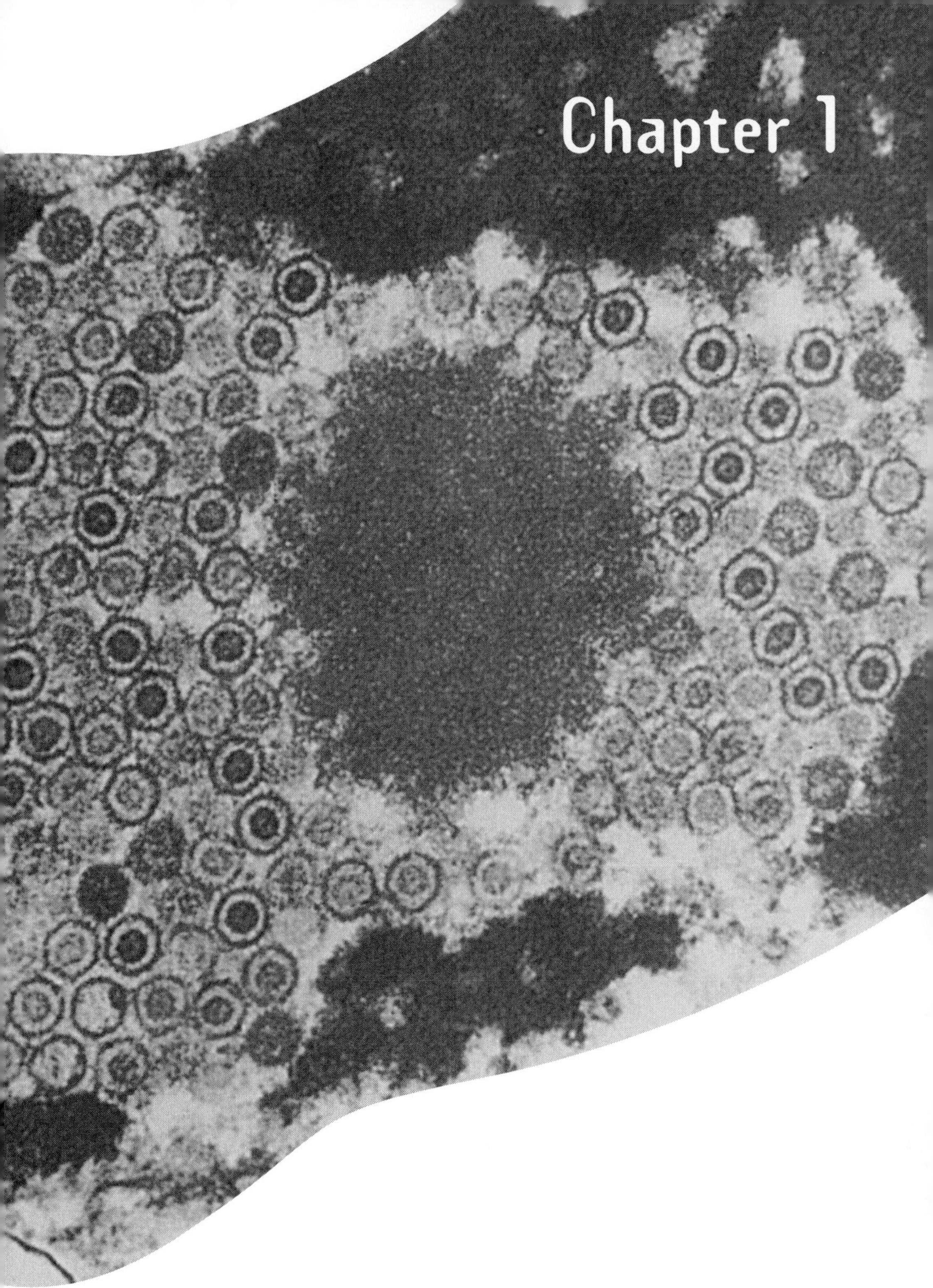

Chapter Opener One is a highly magnified photo of *Herpesvirus*.

Case 1.1

Kate, your sister-in-law, is about to undergo fertility treatments. Her doctor insists that she receive the rubella vaccination, and then wait several weeks before beginning the actual fertility regimen. Kate calls you and wants to know why she has to do this. You ask her if she is able to produce evidence of vaccination for rubella (also known as German measles). She says no; her family had a house fire a few years ago and all those records were lost.

"But I had German measles when I was in second grade!" she says. "I remember that I was really sick and missed almost a month of school."

You suggest that she follow her doctor's advice and get the immunization.

1. Why would a fertility specialist recommend the rubella vaccine? Why does he suggest a waiting period after vaccination and before conceiving?

2. When do most children in the United States receive their rubella immunization?

3. Kate suggests that she had rubella in second grade, but the disease she described doesn't sound like rubella to you. Why not?

4. Kate says the doctor gave her the option of having her blood checked for antibodies to the virus, to test her immune status. Would this test be checking for immunoglobulin M or G (IgM or IgG)? Explain your answer.

5. If a physician was checking for a current rubella infection and only had available a test for IgG, how could he or she be certain the infection was a new one?

Case 1.2

In late September a woman brings her 14-year-old daughter, Meg, to the family physician. Meg shows the doctor the back of her thigh where there are pale red, nonraised discolorations. The rash covers a wide area of the thigh and seems to be roughly circular. The center of the circular area appears normal. Meg has no other symptoms, but her mother brought her in because the rash has been present for over three weeks and it seems to be growing.

The doctor questions Meg about possible exposures. Has she worn any new pants lately? Has she been in the woods? Do her joints hurt? Meg reports that she spent the month of August at summer camp in Vermont. She's been wearing mostly shorts and bathing suits for the past two months, none of them new. She doesn't remember any insect bites on her thigh.

1. On the basis of Meg's **oral history,** what is the most likely diagnosis? What would the causative microorganism look like in a Gram stain?

2. How did she most likely acquire her infection?

3. Would the diagnosis be any different if Meg had attended camp in Arizona? Explain.

4. Why does the doctor ask Meg if her joints hurt?

5. How is this infection treated?

6. Meg's mom, upon hearing the **presumptive diagnosis,** declares that Meg will not return to that camp, which she loves and had planned to attend next summer. The doctor suggests that Meg need only take some precautions. How can she protect herself from getting this infection again?

Case 1.3

You have a possible infectious condition that you are embarrassed to discuss with the physician with whom you work. You have worn artificial nails for several months now and noticed that the one on your left ring finger falls off regularly. The real nail underneath has become white and chalky, and the skin around the nail is beginning to have little white lines in it and look a bit chalky, as well.

1. What disease do you suspect? Explain why.

2. What would you suggest be done for a more definitive diagnosis?

3. Can you treat this yourself with an over-the-counter drug, or do you need to see a physician?

4. You see cures for this condition mentioned on TV and on the Internet—do you think they work?

5. What other conditions are caused by **dermatophytes** (*Microsporum, Trichophyton,* and *Epidermophyton*)? What is special about them that makes them capable of thriving in their anatomical **niche** on their hosts?

Case 1.4

You are the school nurse at Willowdale Elementary. This morning Ms. Matthews, one of the first-grade teachers, brings a little girl named Keisha to your office. Her right eye is swollen and bloodshot. The lining of the lower lid is bright red. There is a thick yellow discharge in the corner of the eye.

1. What is the most likely diagnosis, and what is the **etiology**?

2. What sign leads you to believe that the infection is bacterial in origin?

3. What is the treatment for this condition? Elaborate. Is the condition **communicable**?

4. What are some of the eye's natural defenses that help to prevent infections?

5. Are there steps the teacher should take to prevent the spread of this infection in the classroom? If so, discuss them.

Case 1.5

A woman is brought to the emergency department where you are working **triage.** She has an extremely swollen right lower leg. You see what appears to be an old surgical wound in the mid-calf, with rough scar tissue surrounded by purplish-red skin. She is in a lot of pain and her husband speaks for her. He tells you that three weeks ago she had a group of moles removed from that area. It had appeared to heal initially, but three days ago the incision area started looking bigger rather than smaller. She did not return to the physician, hoping the condition would resolve itself. In the past three days the area has begun to swell and become very hot.

You call the attending physician immediately because you know that this is a serious condition.

The patient is sent straight to surgery where the wound is **debrided.** Gram-positive cocci growing in chains are recovered from the wound. She is transferred to intensive care and put on high-dose intravenous antibiotics for the next 18 hours, but the next evening her leg is amputated below the knee. She remains in the hospital for two months following surgery and requires long-term antibiotic therapy and multiple skin grafts on her upper leg.

1. What condition did this patient have? What features suggest that it is not *Clostridium perfringens* gangrene?

2. Why was amputation the best solution for the infection in this case?

3. How is the bacterium transmitted?

4. It seems like we've heard a lot more about this condition in the past few years. Is this just media hype or are more cases occurring? Explain.

Case 1.6 In the News

In the late winter of 1988, pediatricians in big cities around the country started reporting large increases in the numbers of patients they saw with diffuse red rashes and high fevers (greater than 101°F). The rash, usually extending downward from the hairline to the rest of the body, began after a two-week incubation period. The spots were often so close together that the entire involved area appeared red. Sometimes the skin in such an area peeled after a few days. The rash lasted five to six days. Many of the children also suffered from diarrhea.

The age group most affected was preschoolers. This was a change in **epidemiology** for this infection, as previously the disease most often struck school-age children. A vaccine had been introduced for this disease in 1963, and since then only 5000–6000 cases a year had been reported in the United States. In 1989, 18,193 cases were reported. In 1990 the **epidemic** peaked with almost 28,000 cases reported in the United States. Since then the incidence in this country has fallen rapidly and is again in the range of 5000–6000 cases a year.

1. What was this resurgent infection?

2. What are some possible reasons for the epidemic in 1989–1991?

3. What is **herd immunity**? Discuss it in relation to this outbreak.

4. What is the schedule for vaccination for this infection in the United States?

5. Are serious **sequelae** associated with this infection? If so, what are they?

Case 1.7 Challenge

A woman brings her 6-month-old son to the pediatrician. You are following the doctor as part of your physician's assistant training. Before you enter the examining room the physician pulls the chart off the door and hands it to you. The nurse has written on the chart that the chief complaint is a group of lesions on the child's back.

You enter the room and greet the mother. A toddler girl is leaning over to play with the baby in his carrier on the floor. The baby is giggling and appears healthy. You notice on the chart that the baby was breast-fed from birth through his fourth month. Mom explains that the spots on the baby's back just popped up two days ago and that the baby hasn't had a fever and seems well. She lifts the baby up and you examine the lesions—a group of about seven to eight blisterlike lesions **localized** to the left of the baby's spine. They have clear fluid in them. The physician says the lesions are diagnostic.

1. What are the lesions diagnostic of? Explain how you decided.

2. Although this particular condition is somewhat unusual in babies, the lesions indicate that the child must have experienced a common childhood illness earlier. Which one?

3. The mother says that, to her knowledge, the baby has not had this common childhood illness, but that his 3-year-old sister had it four months ago, when the baby was 2 months old. Explain the link between the girl's illness and the baby's condition.

4. What factors probably influenced the fact that the baby did not have symptomatic illness when his sister was experiencing it? And what factors led to the eruption of lesions now?

5. Is this a dangerous condition? Why or why not?

Case 1.8 Challenge

Your stepbrother John is 5 years old. One day he comes to the breakfast table with a bright red face, almost as if he had been slapped. When you look more closely you can see thousands of tiny red bumps on the skin. He has a milder rash on his arms and legs and just a few red bumps on his trunk. He isn't acting sick and doesn't have a fever. He had chicken pox when he was 3 and his immunization schedule is up to date. He sticks his tongue out at you while you're examining his skin and that reminds you to check his throat, which looks normal, no redness. He says his throat hasn't felt sore. Your mom mentions that he has had a runny nose for the last few days, but he hasn't felt ill.

1. Your diagnosis? Why was his throat checked?

2. Can John go to kindergarten today? Why or why not?

3. Is this infection rare? Explain.

4. Are any **sequelae** associated with this infection? If so, name them.

The Usual Suspects

Common microorganisms causing infections on the skin or in the eyes[1,2]

Bacteria

Gram-positive

Propionibacterium species
Pseudomonas aeruginosa
Staphylococcus aureus
Streptococcus pyogenes

Gram-negative

Borrelia burgdorferi
Chlamydia trachomatis
Haemophilus aegyptius
Mycobacterium leprae
Neisseria gonorrhoeae

Viruses

Adenovirus
Coxsackievirus
Echovirus
Herpes simplex virus
Measles virus
Papillomavirus
Human parvovirus B19
Rubella virus
Varicella-zoster virus

Fungi

Candida albicans
Epidermophyton
Microsporum
Sporothrix schenckii
Trichophyton

Protozoa

Leishmania species
Loa loa
Onchocerca volvulus

[1]Not all of the infections appear in this chapter.
[2]Not an exhaustive list.

Diseases of the Nervous System

Chapter 2

Chapter Opener Two is a highly magnified photo of *Neisseria* meningitis.
Courtesy of D.S. Stephens, Emory University School of Medicine.

Case 2.1

A mother brings her baby daughter in to your office for her 12-month set of vaccinations. The baby is scheduled to receive, among others, the **MMR** and the polio immunizations. After the nurse shows the mother, the baby, and the baby's 3-year-old brother to the examining room, she tells the mother to undress the baby except for her diaper. She hands her a blue booklet about the polio vaccine and additional forms about the other vaccines and asks her to read them all.

The baby cries as the mother undresses her. The 3-year-old starts to climb up a chair and reaches for the needle-disposal container mounted on the wall. The young mother is frazzled; she keeps one hand on the baby lying on the table and tries to scoop up her son with the other. The 3-year-old slips out of her grip and bumps his knee on the corner of the table as he skids to the floor. Now he is crying, too.

The mother never manages to read the vaccination information but she does sign the forms where she is supposed to. When the physician arrives, she checks for the signatures, then performs a thorough examination. On her way out she tells the mother that the nurse will be in to administer the vaccinations.

1. What type of information is contained in all vaccine brochures? Why should they be read before the vaccines are administered?

2. What particular facts are critical for parents to know about the polio vaccine?

3. What other vaccines besides the MMR and polio are appropriate for a 1-year-old child?

4. Two forms of the polio vaccine are available, the live attenuated version (called the OPV, or oral polio vaccine) and the killed or inactivated version (called the IPV, or injectable polio vaccine). Why is the OPV the preferred version for this age?

Case 2.2

You are doing a rotation in the hospital's clinical laboratory. A sample of cloudy cerebrospinal fluid (CSF) from a suspected meningitis case arrives and you are told to Gram stain it, and then to plate it on blood agar and **chocolate agar.** In the Gram stain you find gram-negative rods of varying size and shape. You also find a lot of bacteria inside phagocytic cells. They are not diplococci. Colonies grow on both of the plates you inoculated.

Later, the charge nurse tells you that the patient, a 3-year-old girl, has not received any childhood vaccinations.

1. What is the most likely causative organism? Why?

2. Why was the child's unvaccinated status helpful in diagnosis?

3. What is causing the cloudiness in the CSF?

4. What other types of infections can this organism cause in children?

Case 2.3

You are working in the emergency department of a regional hospital in rural Kentucky. A patient is brought in by emergency medical technicians (EMTs). Their initial report is suspected meningitis because the patient has a headache and stiff neck. The EMTs add that the patient's meningitis symptoms appear rather mild—he still has neck movement, and the headaches are not severe. The patient's overall condition is poor, however. He is very thin, has dark spots on his face and upper body, and open bloody-looking eruptions on his lips. His fever is 104°F, and his blood pressure is low. He also has severe diarrhea.

1. What is the first step in determining if the patient has meningitis?

2. This test reveals the presence of very large cells that appear to be eukaryotic, surrounded by a large capsule. What is the probable diagnosis? Name some other eukaryotic organisms that can cause meningitis symptoms.

3. What groups of people are at risk for this infection?

4. How is it acquired?

5. What anatomical sites are most often infected with this fungus?

6. Let's say your initial suspicion (your answer to question 2) was correct. What other diagnostic test should be performed on this patient?

Case 2.4

You are an emergency medical technician and are called to the home of Kevin, a 13-week-old boy who has become listless and is having trouble breathing. The parents report that Kevin used to smile, but lately he has not smiled, nor has he had other noticeable facial expressions in the last two days. Kevin's eyes are open when you arrive, but he does not seem to be focusing. You place your out-stretched finger under his fingers and he fails to grasp it. You lift his foot and it drops back to the mattress. The parents report that he has not had a bowel movement in three to four days.

1. What is your suspicion, based on what seem to be nervous system symptoms?

2. If this is indeed the case, do you start treatment here at Kevin's home, or should you transport him to the local hospital?

3. What should be administered to Kevin at the earliest opportunity?

4. How do babies acquire this condition?

5. Although the diagnosis should be confirmed with laboratory tests, the tests should probably not be performed in the hospital lab. Why not?

Case 2.5 In the News

In the winter of 1993, five students from one middle school in Seattle were diagnosed with meningococcal disease. The incidence of the disease had been climbing for two years in that area of Washington State, as well as in the rest of the country, and has continued to climb since then. In the Seattle outbreak, health officials identified one strain of the causative organism that was responsible for the increased incidence.

1. What type of organism would you look for in a Gram stain of blood or cerebrospinal fluid in these cases?

2. What is the organism's **portal of entry** to the host?

3. Could you swab the portal of entry (see question 2) to detect the presence of the organism? Why or why not?

4. What types of symptoms are associated with meningococcal disease?

5. A total of 900 students attend the affected middle school. What measures should have been taken to protect the remaining 895 students from acquiring meningococcal disease?

Case 2.6 In the News

In late July of 2000 the most famous park in the United States, Central Park in New York City, was closed to the public so that it could be sprayed with insecticide to prevent the spread of the West Nile virus. Parks department workers handed out pamphlets titled *Public Health Alert.*

West Nile virus had first been noticed in New York the previous summer and fall. Seven people were killed in that early outbreak, and 55 cases of the illness were confirmed. It had not been seen previously in the United States. As its name suggests, it is normally found in Africa, the Middle East, and western Asia, as well as in parts of Europe.

1. How is West Nile virus transmitted?

2. The virus had another vertebrate host besides humans when it showed up in New York. What was it?

3. Can you list some possible mechanisms for how the virus was introduced into the United States?

4. Most infections result in no noticeable symptoms. Some of those infected may develop a skin rash. A fraction of people infected develop life-threatening encephalitis. What is encephalitis and who do you suppose is most likely to experience this symptom?

5. A sudden increase in a particular disease within a population of humans is called an **epidemic.** What is a large outbreak among animals called?

6. If you lived near Central Park and wanted to go jogging there, what would be the best time of day to avoid the park to minimize your chances of being infected with West Nile virus?

Case 2.7 In the News

In the late 1980s there was an **epidemic** among livestock in Great Britain. Approximately 180,000 cattle were found to have "mad cow disease," so named because the condition attacks the central nervous system, which leads to bizarre behavioral symptoms and often death. The disease seems to be caused by an "unconventional transmissible agent," meaning it is unlike most microorganisms. No genetic material from this organism has ever been detected in infected tissues, although foreign protein fibers accumulate in large concentrations in the brain. Infected cows were still turning up in the late 1990s.

There was great alarm in the late 1990s when dozens of humans starting turning up with symptoms similar to those seen in the cows in the late 1980s. More than 50 people have been diagnosed with the human variant of mad cow disease. This is consistent with the approximately 10-year incubation period of this unconventional transmissible agent. In 1996 scientists confirmed that the same agent was present in affected human and cow brains.

1. What is the name for a transmissible agent that contains only protein and has no genetic material?

2. What is the formal name for mad cow disease? Explain the name.

3. The human form of the disease is called something else. What is it?

4. Scientists suspect that the humans infected during this outbreak acquired the disease from eating meat from diseased animals. Even when meat is well cooked, it transmits the infection. What does this say about the infectious agent?

5. These cases in Britain were not the first cases of the disease; it occurs at a low constant rate in other countries, including the United States. Although some of these sporadic cases can be traced to transplants of infected tissues, such as corneas or brain tissues, most are idiopathic. What does *idiopathic* mean?

6. Livestock control measures have been in place in Britain for several years now. Can we expect more human cases with links to the British cattle epidemic, or is it behind us? Defend your answer.

Case 2.8 Challenge

Immediately after you finished the physician's assistant program you took a job at a free clinic in the heart of the city. You always wanted to help society, especially those in the direst of life's circumstances. During your first week, you met Dwight, a 53-year-old overweight white man with several obvious problems. Dwight's feet are cracked and blistered and he has three infected toenails. (He tells you he has been homeless for various periods during the last 10 years.) There are seeping sores in the folds of his wrists and under his arms. His gums are bleeding. Dwight is here because he has been greatly confused over the past eight months and it seems to be getting worse. He has episodes of ranting and raving. He reports "feeling crazy" and being very scared. His friends brought him to the clinic and are waiting outside.

The supervising physician quickly joins you in the examining room. Together you examine Dwight thoroughly and make a plan for addressing his pressing needs for wound care. The physician starts asking him about his health history. Dwight reveals very little, saying his memory is very bad. But he talks a lot about his past sexual exploits.

1. The doctor speaks with you in the hallway. He tells you that mental illness is very common among the homeless population. Dwight needs a thorough psychiatric evaluation. The doctor is fairly sure that some of Dwight's neural symptoms are caused by a sexually transmitted infection. Which one? Caused by which microorganism?

2. If the blood test comes back positive, does it mean that Dwight can transmit the disease to others? Explain.

3. Should Dwight be treated with antibiotics to remedy his neural symptoms? Why or why not?

4. The patient's blood test came back positive. For what other infectious disease should he now be tested?

5. The doctor tells you to expect to see more of these cases in the future. But a coworker, who graduated from nursing school 10 years ago, tells you that this disease (especially its later forms) is relatively rare and is decreasing in incidence. Who is right?

The Usual Suspects

Common microorganisms causing infections in the nervous system[1,2,3]

Bacteria

Gram-positive

Clostridium botulinum
Clostridium tetani
Listeria monocytogenes
Streptococcus pneumoniae

Gram-negative

Haemophilus influenzae
Mycobacterium leprae
Neisseria meningitidis
Treponema pallidum

Viruses

Measles virus
Poliovirus
Rabies virus
Varicella-zoster virus
West Nile virus
Arboviruses (western equine encephalitis, eastern equine encephalitis, St. Louis encephalitis, etc.)

Fungi

Cryptococcus neoformans

Protozoa

Naegleria fowleria
Toxoplasma gondii
Trypanosoma species

[1]Not all of the infections appear in this chapter.
[2]Not an exhaustive list.
[3]Prions also cause disease in the nervous system.

Diseases of the Cardiovascular System

Chapter 3

Chapter Opener Three is a highly magnified photo of *Pneumocystis carinii*.
Courtesy A.B Dowsett/Photo Researchers, Inc.

Case 3.1

You are at your son's baseball game when another boy's dad experiences dizziness and nearly faints in the stands next to you. You tell him that you are a paramedic and will walk him to your car where you have your medical equipment. He reports that he has had a headache off and on since he had a tooth extracted four days ago. This evening he is feeling very weak.

His blood pressure is normal. When you listen to his heart you note that he has a pronounced **murmur**. He reports having had rheumatic fever 15 years ago. You examine his fingernails and find one that has tiny **petechial** hemorrhages under it.

1. Which cardiovascular infectious condition is this?

2. What is the most likely causative organism and the route of transmission?

3. What's the connection, if any, with rheumatic fever?

4. Why did you look at his fingernails?

5. What type of culture would a physician most likely order, and why?

6. What is the treatment? Is there a way to prevent the condition?

Case 3.2

On Christmas Eve, 2000, you were working as a clerk in a Dallas emergency room. At 3 A.M., a man and two women arrived with a screaming 6-year-old girl. The man tried to explain what was wrong, but he spoke only Spanish and you had a difficult time understanding him. The girl's mother was sobbing and you couldn't hear what she was saying. The other woman spoke a bit of English and explained to you and the nurse on duty that this was her sister's family, who had just arrived from El Salvador. The aunt did not know what was wrong with her niece but told you that the father was repeating the words for "break bone."

The nurse examined the girl and found that she had a rash and a fever of 104°F. Although the girl seemed to be in severe pain, the nurse found no bone fractures. The father, shaking his head violently, said something urgently to his sister-in-law. She interpreted his frantic statement for you, "He said it's in her blood."

1. What is your diagnosis?

2. What connection does this disease have to broken bones?

3. This is a vector-borne disease. What is the name of the most common vector?

4. What other infection is transmitted by the same vector?

5. The next night when you arrived at work the little girl seemed to be doing better. The rash had subsided and her fever had lowered. But on the third night you arrived to find that she had been transferred to intensive care after hemorrhaging internally. Is this still consistent with your original diagnosis? Explain.

6. This all sounds very bad, but you're somewhat comforted by the fact that this disease is not found in the United States. Right?

Case 3.3

You work in a small family practice in rural Virginia. A man in his early 50s comes in with a complaint of intermittent fever (102–103°F) and headache for the past two weeks. The physician examines him and takes a history. The only clinical finding is a wound about the size of a quarter on his right thumb. **Axillary** lymph nodes are swollen and tender. The man says he cut himself while skinning a rabbit three days ago. On the basis of these observations the physician prescribes streptomycin and asks the man to call if his symptoms don't improve in three days.

The physician asks you to draw blood and tells the patient he should return in four weeks for another blood sample. She says there is no need to culture the wound.

1. On the basis of the limited information above, the physician has obviously made a diagnosis. What is it? What does it look like when Gram stained?

2. What is the most likely reservoir for the causative organism in this case?

3. Why draw blood twice?

4. Why not culture the wound to look for the bacterium?

5. What are some other common infections that humans acquire from animals? (These are also known as **zoonoses.**)

Case 3.4

A 63-year-old international telecommunications executive visits your office with complaints of a high fever. The fever is not constant, but **intermittent.** When you press him for details he estimates that every three days or so he suffers these **debilitating** "sweats." He usually has headaches and muscle aches during the episodes. They keep him home from work. After half a day or so he feels better. He reports that he has experienced these episodes for about two months.

1. What is the name of the condition you suspect?

2. What should be your first question about the patient's history?

3. What is the most likely causative organism (genus and species)? Support your answer.

4. Is this pathogen eukaryotic or prokaryotic?

5. Which is the most dangerous of the species that can cause this disease? Give some details.

6. What are the two main places in the human body that are exploited by the causative organism in this disease?

7. Can this individual transmit this infection to others? Why or why not?

Case 3.5

You've decided to work in the Peace Corps for the first two years after graduating from nursing school. Your assignment is in a rural area in South Africa. You and a coworker are setting up a clinic and encouraging women from the surrounding villages to bring their children when they are ill and to visit the clinic themselves, especially when they are pregnant.

1. In your first week you saw several children whose major symptoms were high fever, lots of sweating, and **prostration.** They all turned out to have the same infectious condition, one that you continued to see throughout your stay in South Africa. Up to half of the sick children did not survive this illness. What is it?

2. In this setting, what is the best prevention for this disease?

3. In your third month you saw a 2-year-old boy with an angry-looking rash. He was very ill with a high fever, and eventually died. His death surprised you because you thought this disease had been conquered long ago. (In the United States it is seen only occasionally, because children are vaccinated for it.) Over the course of your two-year stay you saw these symptoms in children perhaps a dozen times. Several of the children died. What is the disease?

4. Name at least two of the most common infectious conditions you should look for in adult clients in this setting.

Case 3.6 In the News

A newspaper report from Boston in the late 1990s described a growing fear among local residents. They were afraid to venture outdoors because of the increasing visibility of a particular infectious disease. The article reported that the number of people hiking in Massachusetts had recently decreased dramatically, and that many homeowners were erecting fences and spraying their yards with pesticides. Many people who dared to venture outdoors wore white clothing and tucked their pants inside their socks. In New York, there were reports of residents simply paving over their lawns, and some people gave up gardening altogether.

Although most prominent in the Northeast, similar behaviors were seen all over the country. In Montana, 10% of people surveyed felt they were at high risk for the disease, even though the Centers for Disease Control and Prevention (CDC) said that the risk was very low in that state.

1. What infectious disease do you suppose these cautious citizens were trying to avoid?

2. What determines which geographical region of the country carries risk of this disease for its inhabitants?

3. Another major disease in the United States is transmitted in a similar way. What is it, and what microorganism causes it?

4. Which regions of the country have a high incidence of this second tick-borne disease?

5. Which of these two diseases frequently has no skin manifestations at all?

Case 3.7 Challenge

Fred is a longtime patient in the family practice where you work. Typically he comes in once a year for a physical because his job involves high-steel construction work and his company requires annual checkups. However, during the past six months he has visited the office three times.

Fred first came to the office in January complaining of extreme fatigue. He had lost 15 pounds since his checkup the previous May. As part of his examination the physician ordered a standard human immunodeficiency virus (HIV) test (of the ELISA type), which came back negative. He was to come back for a repeat test in May. But he returned one month later because he was experiencing an episode of genital herpes in which the lesions had not healed in over three weeks.

The episode eventually subsided and Fred returned in May for his repeat HIV test. During this visit, he told the doctor that he was recovering from a severe respiratory infection that had bothered him for weeks. The physician drew blood; his **CD4 count** was 200 cells/ml. This HIV test (again, an ELISA) was also negative.

Two months later Fred was admitted to the hospital and a lung biopsy demonstrated *Pneumocystis carinii* pneumonia, but another HIV test came back negative. He was released after three weeks, but re-admitted with the same infection two months later. Again he tested negative for HIV. He died three days after admission to the hospital.

1. What is an ELISA test, and what does the one for HIV actually detect?

2. This patient did indeed have HIV infection, but continued to test negative. What are some possible explanations for the consistently negative test results?

3. Are any alternative tests available to clinicians for patients strongly suspected to be HIV-positive who test negative with the usual test?

4. Would you expect patients with lack of serum reactivity to have a fast or slow progression from HIV infection to acquired immune deficiency syndrome (AIDS)? Defend your answer.

5. Which of the reported symptoms are consistent with a diagnosis of HIV?

Case 3.8 Challenge

You're at the beach on Lake Michigan with your friends over spring break. The house you're staying in is a few blocks away from the beach (okay—so you're on a budget!), and the flower border around the house is overgrown with weeds. There is a tiny concrete patio next to the house where the four of you crowd to lie out in the sun when you're not at the beach.

Everything is fine until Janet complains of an insect bite on her ankle. It looks like a big mosquito bite. You rummage around under the sink in the bathroom and find a very old bottle of aloe lotion. She rubs it on the bite and you both return to the patio.

The next day Janet's ankle is very red in the area around the bite. It is hot and tender to the touch. Being nursing students, you decide not to take a chance and you drive her to the local hospital's emergency department to have it looked at. You wait there for four hours while other, more seriously ill patients are seen before you. It's your last day at the beach, and even Janet is beginning to feel it's not worth wasting the day in the waiting room. So you leave the hospital without seeing a doctor.

You go back to the house and Janet puts more aloe lotion on the bite. Then off you go to the lake. That night Janet's roommate wakes you at 2 A.M. saying that Janet is crying and sweating. When you get to her room you see that Janet looks very ill. She is covered in sweat but is shivering. She is very pale, almost blue in places, and there are red patches on her legs. You dial 911.

1. What do you suppose is happening with Janet? Is it dangerous?

2. Explain Janet's symptoms described in the last paragraph of the case.

3. What organism causes this condition?

4. When you relate the history of Janet's condition to one of the paramedics, you notice that she writes "secondary to cellulitis" on her pad of paper. What is cellulitis, and what does it mean that Janet's condition is "secondary" to it?

5. How should Janet's condition be treated at this point?

The Usual Suspects

Common microorganisms causing infections in the cardiovascular system[1, 2, 3]

Bacteria[3]

Gram-positive

Bacillus anthracis
Clostridium perfringens
Streptococcus pyogenes
Other streptococci

Gram-negative

Bartonella henselae
Borrelia burgdorferi
Brucella species
Ehrlichia species
Francisella tularensis
Rickettsia species
Yersinia pestis

Fungi

Various

Viruses

Coxsackievirus
Dengue fever virus
Ebola virus
Epstein-Barr virus
Yellow fever virus
Human immunodeficiency virus

Protozoa

Babesia microti
Plasmodium species
Schistosoma species
Trypanosoma cruzi
Wucheria bancrofti

[1]Not all of the infections appear in this chapter.
[2]Not an exhaustive list.
[3]Many bacteria can cause bloodstream infections if given access; this table lists those adapted to cause disease in this system.

Diseases of the Respiratory System

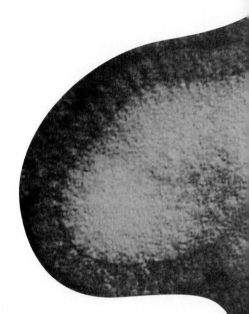

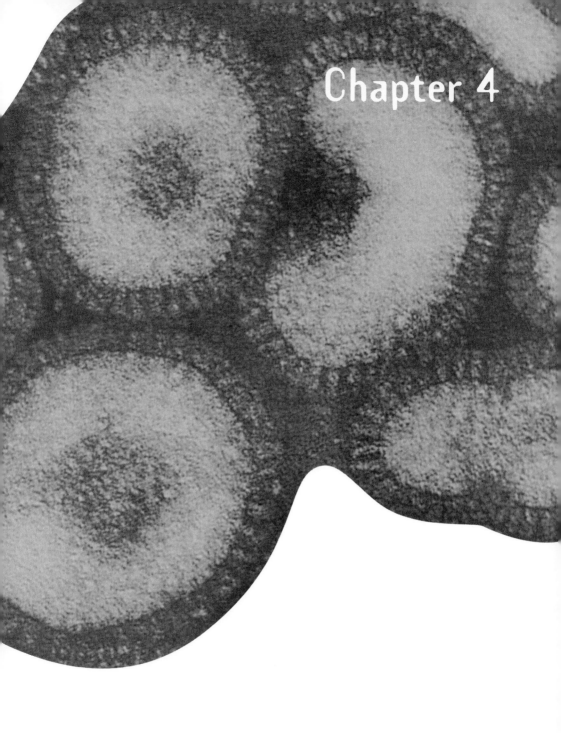

Chapter 4

Chapter Opener Four is a highly magnified photo of *Influenzavirus*.
Courtesy G. Murti/Visuals Unlimited.

Case 4.1

You are a physician's assistant at a local pediatrician's office. Five-year-old Michael is brought to the office by his father. Michael is crying and complaining that his mouth hurts. His father has been at work and does not know whether the boy has had a fever during the day. Currently his temperature is 103°F. The physician notices that Michael's breath smells rotten. Lymph nodes in his neck are swollen, and visual examination of the throat reveals a white packet adhering to the left tonsil. Much of the soft palate is red.

1. What laboratory tests are called for?

2. What types of infections are in the **differential diagnosis**?

3. Your practice has recently been overrun by sore throats and now, late in the evening, there are no supplies for performing the proper test. Should the physician prescribe antibiotics or not?

4. In deciding whether to prescribe antibiotics, should the physician be extra careful not to prescribe an unnecessary antibiotic, or be extra careful not to let a bacterial infection go untreated?

5. What are the possible **sequelae** of untreated sore throats?

Case 4.2

You and your friends are driving to the mall; it is late October. A public service announcement comes on the radio urging people to get their flu vaccinations. You are a second-year nursing student and you mention that the nursing staff at your university is holding a vaccine clinic next week.

Your friend Susan says, "I'm not getting a flu shot! Last time I did, it gave me the flu." Others in the car agree with her. But Heather asks you if it's true that the vaccine can give you the flu.

1. What should your answer to Heather's question be?

2. Heather says that because she had a flu shot last year she's going to skip it this year. Respond, with an explanation.

3. What is the difference between "**antigenic** drift" and "**antigenic** shift"?

4. What is different about the vaccine from year to year? Who decides what form it will take every year?

5. Susan wants to know why you don't have to get other vaccines annually.

6. Another friend, Dru, says that even though she had the flu shot last year she got terribly sick with the stomach flu over Thanksgiving break and missed most of her vacation. What is your explanation for this?

Case 4.3

Julie's husband Doug has not been feeling well for the past 10 days. He has congestion in his lungs and has been very tired. She talked him into going to the doctor a week ago when his temperature was 101°F. The doctor gave him some oral amoxicillin, which he took faithfully until it was gone. But she still thinks he looks sick.

Julie, Doug, and their 3-year-old daughter have just moved to Ohio from Arizona. Doug is a park ranger and he loves his job, but for the past three days he has felt too sick to go to work. His respiratory symptoms have not improved. Julie makes an appointment for him with her doctor.

1. As the physician's assistant in the office, you are the first to examine Doug. What's your tentative diagnosis, based on the history?

2. Which components of the history support your tentative diagnosis?

3. Doug's condition has not responded to the antibiotic. List two possible reasons for this finding.

4. What are some other conditions caused by this microorganism?

5. Should Julie worry that Doug can transmit the infection to her or to their daughter?

6. What precautions can be taken by other workers who may be regularly or heavily exposed to bat or bird droppings?

Case 4.4

It is mid-July. You are working as a **triage** nurse in the emergency department of a small suburban hospital in Arizona. A young, athletic-looking man in his early 20s is helped into your office by his girlfriend. He greets you and sits down, but is feverish and his breathing is labored. The girlfriend answers your questions for him. She says the symptoms began about 24 hours ago and seemed to worsen quickly. It looks like the flu to you, but the season is wrong. So you ask about the man's activities over the past week to 10 days. Nothing in this history points to an obvious **etiology** for the disease. And the girlfriend, rather defensively, adds that she is a "neat-freak" and is constantly cleaning and disinfecting the house they share. But of course, respiratory infections are very common and can be acquired anywhere. After listening to his chest you decide that it may be bronchitis or influenza. You decide to isolate him from the rest of the people in the waiting room until an examining room becomes free.

Forty-five minutes later the girlfriend comes barreling into your office. "I think he's choking!" she screams. You and the attending physician arrive at his bed where he indeed seems to be suffocating. His face is red and he is gripping his throat. The doctor calls out, "**acute** respiratory distress," and a team moves in to try to restore his breathing.

Later that evening, on your way out, you learn that the patient has died. Several days later the charge nurse tells you what the patient's lab work revealed. It identified an infection that he probably acquired a few weeks earlier while he and his girlfriend stayed in an isolated cabin his family owned but seldom used.

1. What is the diagnosis?

2. What connection does the diagnosis have with the cabin?

3. You overhear the charge nurse say to herself: "I knew there was a good reason not to clean my house." To what could she have been referring?

4. This case is from Arizona. These infections were first seen in the United States in May of 1993 in the Four Corners area of the Southwest, which includes Arizona, Colorado, New Mexico, and Utah. Can we assume that this disease is only found in the Southwest? What factors determine the places this virus might be present?

Case 4.5

Your son's best friend, Josh, has infectious mononucleosis; he hasn't been in school for two weeks. Your son and three of his friends come over after basketball practice looking for snacks, but they also want to talk to you about Josh's infection because they know you are a physician's assistant. They are all afraid to visit Josh, but they want to know when they can expect him back at practice. One of the boys asks you what causes "mono." Another one of the boys says he heard it was a form of herpes. All of the boys cringe at that one. Can you help these guys out with some information?

1. What causes mono, or infectious mononucleosis? What do you know about this agent?

2. What are the symptoms?

3. How long will Josh be out of school? Is it okay to visit him?

4. You tease the boys by saying, "Besides, by the time you're adults, all of you will have it anyway." Before they recover from that shock you add, "and some of you have it right now!" Are you just playing around with them, or are these statements true? Explain your answers.

5. Sam, the point guard on the team, says his aunt has **chronic** fatigue syndrome. "Isn't that caused by the same virus?" he asks. Is it?

Case 4.6

When you left for school this morning your 3-month-old son was wheezing a bit and he had a slight fever of 99.8°F. Your mother is watching him while you come to school to take your anatomy and physiology exam. Your pager goes off halfway through the exam. The baby's fever is rising and he is having more trouble breathing. Your mother says she is taking him to the emergency room. You rush over to the hospital. When you get there, he is in an examining room and the doctor is signing papers to admit him to intensive care. She says she suspects some kind of pneumonia. She mentions the type of pneumonia but you don't recognize the name and you are too worried about your son to pin her down at this moment. You do note that she mentions that the hospital has seen a dozen pediatric cases of this same type of pneumonia in the past week and a half.

The doctor swabs your son's nose but says the results won't be back for several days. In the meantime, they will give him supportive therapy, including an inhaled spray, but no antibacterial drugs. The doctor says that she feels sure that the child will recover, since the infection was caught very early. Nonetheless, after she leaves, your mother is frantic and indignant. She fires off the following questions to you.

1. What kind of pneumonia is it?

2. Why aren't they giving him antibacterial drugs?

3. How can the doctor be sure what's causing the pneumonia if she doesn't yet have test results?

4. What about your other child, who is 3 years old? Has she been exposed to the infection by being around the baby? Should the baby remain isolated when he comes home? Can the 3-year-old be vaccinated?

Case 4.7 In the News

One autumn in the late 1990s, a number of people became ill after working at a single building at an industrial plant in a neighborhood of Baltimore, Maryland. Their symptoms ranged from simple coughing and other respiratory symptoms to pneumonia. At least one of the 70 people reporting symptoms died.

The company voluntarily closed the building upon the recommendation of the Maryland Department of Health and Mental Hygiene. After all of the water systems at the plant were evaluated and disinfected, it reopened and no new cases were reported.

1. Health departments often have even less information than this when they have to start hypothesizing about the causative organism and its source. What is your first guess?

2. Describe the transmission characteristics of the suspected bacterium.

3. Is there a risk for a continuing community outbreak from these initial infections? Why or why not?

4. Would the health department be likely to identify this bacterium by performing routine water-screening procedures, such as serial dilution or filter collection followed by incubating on eosin methylene blue (EMB) or nutrient agar? Explain.

Case 4.8 Challenge

You have just been accepted into the nursing school at a local medical center. The program requires that you have a physical, which includes a tuberculosis (TB) test as well as the hepatitis B **recombinant** vaccine series. The nurse administering the TB skin test explains that if significant swelling occurs around the injection site, you will probably have to have a chest X-ray to determine if you are infected with *Mycobacterium tuberculosis*. One and a half days later you wake up and look at your arm, which appears swollen in an area about the size of a quarter around the skin test. It is red and tender to the touch. You're alarmed; could you have TB?

1. Why does the reaction take 36–48 hours to show up?

2. If you have a tuberculosis infection, why doesn't the whole body, or at least the respiratory tract, react when the **antigen** is injected during this diagnostic test?

3. You are referred for a chest X-ray, but the results are inconclusive. The clinic doctor prescribes a six-month course of isoniazid (abbreviated INH). You take the medicine according to the pharmacist's instructions. Six months later you are taking a medical microbiology course as part of your nursing curriculum. On the day you study tuberculosis, you suddenly realize why you had a positive skin test. It has nothing to do with a true infection, but with the fact that you were born in the Netherlands. Your family moved to the United States when you were 4 years old. What do you suppose is going on here? Discuss as fully as you can.

4. You have a friend in your hometown who is HIV-positive. When you told her about your TB scare, she said that her specialist can't use the TB skin test, even though HIV-positive people are at higher risk than the healthy population for TB. Why is the skin test not recommended for HIV-positive people?

The Usual Suspects

Common microorganisms causing infections in the respiratory tract[1,2]

Bacteria

Gram-positive

Corynebacterium diphtheriae
Stapyhlococcus aureus
Streptococcus pneumoniae
Streptococcus pyogenes

Gram-negative

Bordetella pertussis
Chlamydia pneumoniae
Coxiella burnetii
Haemophilus influenzae
Legionella pneumophila
Mycobacterium tuberculosis
Mycoplasma pneumoniae
Yersinia pestis

Viruses

Coronavirus
Epstein-Barr virus
Hantavirus
Influenza virus
Parainfluenza virus
Respiratory syncytial virus
Rhinovirus

Fungi

Aspergillus
Blastomyces dermatitidis
Coccidioides immitis
Cryptococcus neoformans
Histoplasma capsulatum

Protozoa

Uncommon

[1]Not all of the infections appear in this chapter.
[2]Not an exhaustive list.

Diseases of the Digestive System

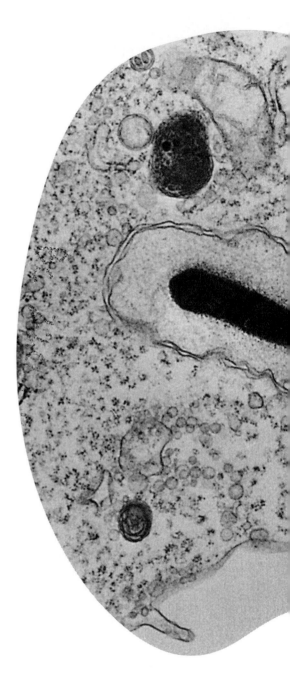

Chapter 5

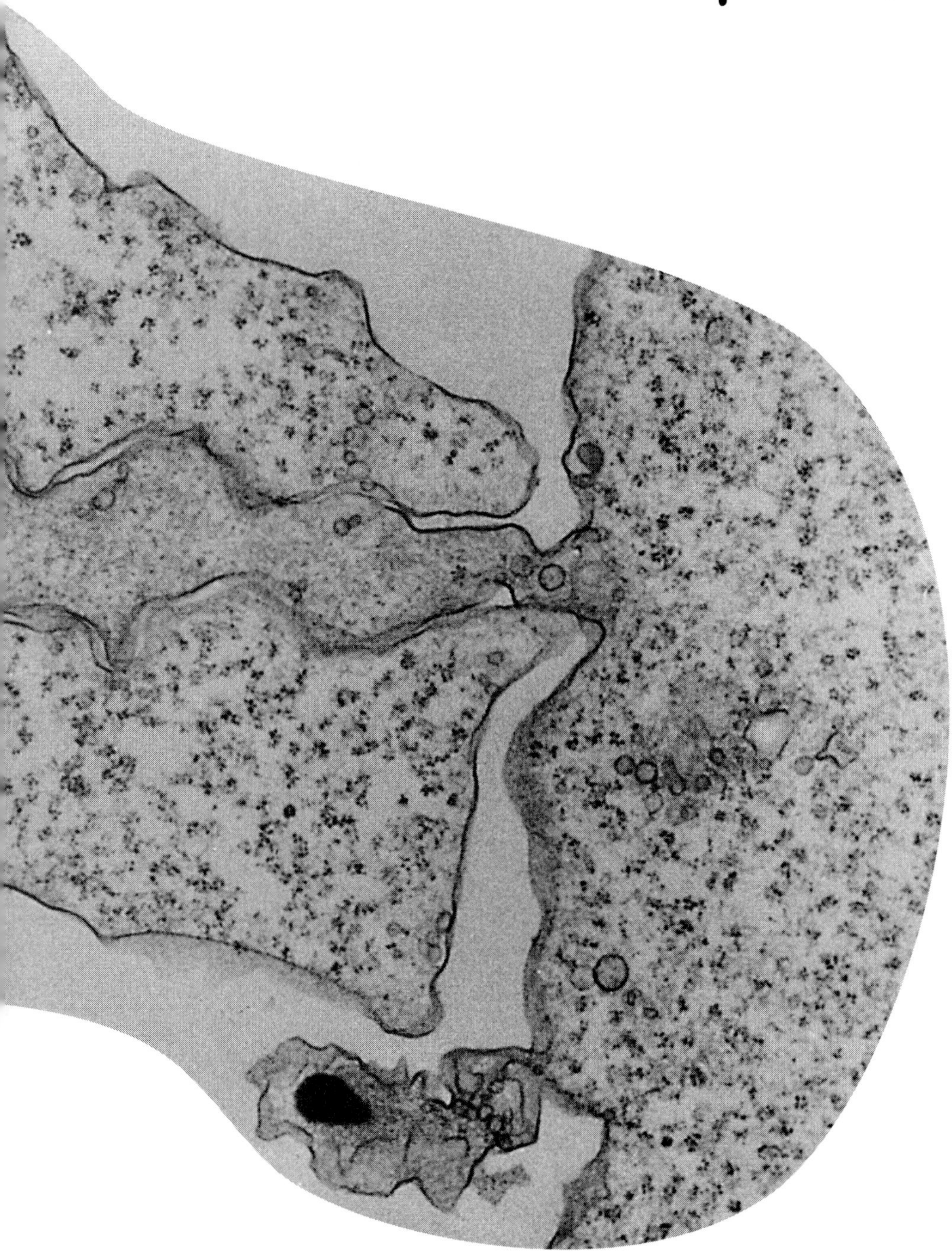

Chapter Opener Five is a highly magnified photo of *Listeria monocytogenes*.
Courtesy of L. T. Tilney, P.S. Connelly & D.A. Portnoy.

Case 5.1

You are at dinner with four of your friends. A local outbreak of *Escherichia coli* O157:H7 has been in the news. The news stories suggest that the source of the infection was unpasteurized apple cider, but the group wants to know if hamburgers are safe. They remember that there was a big outbreak of *E. coli* associated with burgers from a fast-food restaurant in the Northwest. They turn to you, since you are a nurse. You tell them to order steaks. They ask if you're buying!

1. Why steaks instead of hamburgers?

2. One of your friends acts disgusted and says she'll order a salad instead. Will this guarantee her safety? Why or why not?

3. One of your friends says that her sister gives her baby apple juice every day. Should she stop? Explain your answer.

4. What are the symptoms of *E. coli* O157:H7 infection?

5. Another friend says that his family has always eaten rare hamburgers and no one has ever gotten sick. He thinks it's all a bunch of overblown media coverage and says he will continue to eat his favorite delicacy, raw hamburger meat on crackers. What should you tell him?

Case 5.2

Last week you were on a clinical rotation at the local hospital as part of your second-year nursing program. On this rotation, your instructor took a hands-off approach and left you on your own for hours at a time. You spent most of your time hanging around at the nursing station, following nurses as they went about their duties from bed to bed, and listening to conversations between doctors and nurses about patients.

Then, one day one of the nurses who had just emerged from his fourth trip to the bathroom collapsed behind his desk. He had been losing weight and today looked especially pale. You ran to get the attending physician who was just across the hall. He took one look at the prostrate nurse and said something like "see dif" to the nursing instructor who had arrived on the scene. She replied that he had been on multiple antibiotics for the past few months in an attempt to treat a particularly nasty sinus infection.

After the sick nurse is transferred to a bed, your instructor asks you for a written report on the condition. You didn't want to admit that you weren't really sure what condition was involved here, so you figured you could look it up in your books or on the Internet at home.

1. Your Internet search of all kinds of different spellings of "see dif" yields nothing. What section of your microbiology text would likely contain the help you need? What clues lead you in that direction?

2. Now that you've found the right category of infections, can you identify what "see dif" is?

3. Your book has only a small paragraph on this infection. But now you know what to search for on the Internet to find more information. Your instructor wants you to report on the **epidemiology** of the infection. You find that it is referred to as an opportunist and this accounts for its epidemiological patterns. First of all, what is an opportunist?

4. Part of an epidemiological description of an infection involves knowing who is most often affected by it. Let's consider opportunistic infections as a group. People in which age groups are most likely to suffer symptoms from an opportunistic infection?

5. In this case the affected nurse is in his mid-30s. Is it his age or something else that predisposes him to the infection? Discuss.

6. What is the major **virulence** factor for this microorganism?

Note on using the Internet for research purposes: Always be sure that you use a reliable website, such as the Centers for Disease Control site, www.cdc.gov. You will probably find hundreds of sites from other sources, such as class notes posted by professors from various universities and student reports, as well as information from pharmaceutical companies that are marketing drugs to treat the infection. Your search may even return personal web pages of people who have suffered from the disease. This information may or may not be reliable.

Case 5.3

Your sister Pam called you last night, upset about her recent visit to the pediatrician (she has a 3-year-old son). Actually, she was upset about the discussion she had afterward with her husband, who was adamantly opposed to having their son vaccinated against hepatitis B virus (HBV). Pam called you because the doctor had convinced her that it was necessary, and indeed routine, to vaccinate young children. Her husband believes that hepatitis B is mostly acquired through sexual contact and drug use and that it's ridiculous to vaccinate a 3-year-old. Pam wants your advice before continuing this discussion with her husband.

1. First of all, is Pam's husband correct about transmission of the virus? Elaborate.

2. How severe is this infection for young children?

3. Pam says she'll also remind him that in the last year the newspapers have reported at least three hepatitis outbreaks traced back to restaurants. Respond to her statement.

4. While you're on the phone with Pam, her husband comes home from work. He hears your conversation, and says in a loud voice, "That vaccine is not safe! It's one of those genetically engineered things!" What can you tell Pam about how the vaccine is made, and whether it is safe or not?

Case 5.4 In the News

One summer in the late 1990s, a group of tourists from the United Kingdom became ill after they all stayed at the same hotel in Greece. Epidemiologists conducted surveys among all the people who had stayed at that hotel during the two-and-a-half-week period in which people were reporting their illnesses. They did this in an attempt to determine the cause of the symptoms, which were primarily diarrhea and nausea. They surveyed 239 people; 224 of them reported having been ill while they were still on vacation or shortly after their return. Their diarrheal symptoms lasted 10–15 days.

Seventy of the 224 people who reported illness were classified as having definite cases of gastrointestinal disease. A case was called "definite" when a pathogen could actually be recovered from their stool. Of these, the vast majority tested positive for one particular microorganism.

1. Microscopic analysis of the stool samples revealed the presence of small oval-shaped structures, with defined outer walls and two to four nuclei inside that looked like seeds. What is your diagnosis?

2. What organisms should be included in the **differential diagnosis** of this infection?

3. What feature of the symptoms suggests that the causative organism is not likely to be *Staphylococcus aureus*?

4. Epidemiologists interviewed the patients about their vacation activities and food intake to try to identify the environmental source of the infection. There was no relationship between illness and a person's attending one of the scheduled children's activities at the hotel. Only two types of food available in the dining room seemed to be associated with the illness: raw vegetables and salads. There was also a statistically significant relationship between illness and having consumed orange juice made from a mix (with hotel water). So what was the likely source?

5. Why would an epidemiologist even ask about a person's attendance at children's activities?

6. Are there any symptoms that would help to distinguish this kind of diarrheal illness from others?

Case 5.5 In the News

On Christmas Eve a few years ago, the Ohio State Health Department announced that two elderly people had died during the previous six weeks, apparently after ingesting tainted meat. Ten additional nonfatal cases were reported in the state during this period.

The state epidemiologist was aware of a national outbreak of a disease with the symptoms seen in these cases. The symptoms included fever and muscle aches and often diarrhea and nausea. Occasionally, the central nervous system was affected, resulting in confusion, stiff neck, headache, and convulsion.

The nationwide outbreak affected approximately 40 people, with a particularly high infection rate in pregnant women and a significant number of deaths among fetuses. The Centers for Disease Control and Prevention (CDC) issued a list of people who were at particular risk for the disease. These included pregnant women, newborns, people with weakened immune systems, and the elderly.

The CDC and the U.S. Department of Agriculture suspected prepackaged meats, such as hot dogs and cold cuts, as the source of the outbreak. A recall of meat processed at a particular plant in Michigan was instituted.

1. What is the most likely causative microorganism in this outbreak?

2. Why is this infection associated with processed meats, but usually not with hamburger or cuts of meat including pork, beef, or chicken?

3. Epidemiologists describe a microorganism's **pathogenicity** as the proportion of people who become ill after being exposed to the microorganism. (An infection that is **subclinical** in most people who acquire it is considered to have low pathogenicity.) After considering the types of people at high risk for the disease, would you suppose that this organism has high or low pathogenicity? Explain your answer.

Case 5.6

You are working as a receptionist at the only family practice in a small town in Idaho while you are studying to become a physician's assistant. On a Saturday morning you are the only office worker there when a call comes in from a local church. The congregation is hosting a family that moved to the United States from Peru six weeks ago and is helping them find housing and work. In the meantime, the family is staying at a church-owned house and relying heavily on church members for help negotiating this new country and for translation while their English is still sketchy.

The woman on the phone identifies herself as Leslie, a church member. She seems distraught. She says that the mother of the young family became ill yesterday and seems extremely ill now. Her symptoms started out as stomach cramps and quickly progressed to a very watery diarrhea. You hear moaning in the background and Leslie tells you that the patient is pointing to her calves and crying. You ask Leslie how many stools the sick woman has had in the last 12 hours. She replies that it is almost constant and that the woman can no longer leave her bed at all.

When asked, Leslie says there is no blood in the excreta. It is very clear with lots of little white flecks in it. You put her on hold and run down the hall to the examining room where a physician is doing a well-baby check.

1. When the doctor opens the door you whisper that you think there's a case of _____ on the phone.

2. The doctor's eyes widen and she asks you how you came to that conclusion. What is your reply?

3. Why was the doctor initially dubious about your diagnosis and why does the patient's recent immigrant status convince her that your diagnosis was correct?

4. The doctor asks you to tell Leslie to call 911. The sick woman should be transported to an emergency room right away and the doctor will call ahead and meet her there. What is the first intervention likely to be performed when the patient arrives?

5. The incubation period for this disease is one to four days. Can you think of any way that the young mother could have been infected so recently even though she has been in this country for six weeks?

6. The next day you ask the doctor about the patient's status. She says that currently the patient is receiving a course of the antibiotic ciprofloxacin, though it won't help her. Why won't it help her and why was it prescribed if it won't?

Case 5.7

You went to get a haircut yesterday and your stylist was having a conversation with another stylist in the shop. The second stylist said that her live-in boyfriend of three years just got a blood test and discovered he has hepatitis C. Your stylist shrugged her shoulders and said her boyfriend has never had any symptoms so she wasn't going to worry about it.

1. After the second stylist walks away, your stylist asks you about hepatitis C. Her first question is, "Is it serious?" Answer this question as thoroughly as you can.

2. How is it transmitted?

3. Can she be vaccinated against it?

4. Your stylist has heard of hepatitis A and hepatitis B, but never hepatitis C. Is it new? Explain.

Case 5.8 Challenge

When you arrived at work in the intensive care unit this morning, you learned that a patient with Guillain-Barré syndrome had been admitted. He is a 45-year-old poultry farmer.

He is on a respirator and has **bilateral** paralysis of his legs. You remember that Guillain-Barré syndrome is a result of the immune system attacking the peripheral nervous system.

1. What leads to Guillain-Barré? What would you look for in the patient's history?

2. Considering this patient's profession, what type of condition do you suspect as the **precipitating** event?

3. Is Guillain-Barré often fatal?

4. What about the original infection, which you identified in question 2? Is it common or uncommon?

The Usual Suspects

Common microorganisms causing infections in the digestive tract[1,2]

Bacteria
Gram-positive
Bacillus cereus
Clostridium difficile
Clostridium perfringens
Listeria monocytogenes
Staphylococcus aureus
Streptococcus mutans

Gram-negative
Campylobacter jejuni
Escherichia coli O157:H7
Other *Escherichia coli*
Helicobacter pylori
Salmonella species
Shigella species
Vibrio cholerae
Yersinia enterocolitica

Viruses
Epstein-Barr virus
Hepatitis A
Hepatitis B
Hepatitis C
Hepatitis E
Mumps virus
Rotavirus

Fungi
Candida albicans

Protozoa
Cryptosporidium
Entamoeba histolytica
Giardia lamblia

[1]Not all of the infections appear in this chapter.
[2]Not an exhaustive list.

Diseases of the Genitourinary System

Chapter 6

Chapter Opener Six is a highly magnified photo of *Gardnerella vaginalis*.
Courtesy of the CDC.

As part of a community service requirement in your second-year nursing program, you are volunteering at the local clinic for sexually transmitted diseases (STDs). At the clinic, you are responsible for conducting intake interviews. When patients arrive, they fill out a questionnaire and then you take them to an office to go over their answers with them.

Your first patient is a 22-year-old woman. On the questionnaire, she lists her chief complaint as "pain in her belly." You wonder why she has come to the STD clinic for belly pain and speculate it is because this clinic is the only one in town that is free. You excuse yourself and ask the head nurse if it's okay to continue with this client's questionnaire. "Why wouldn't it be?" is his answer. You say you don't think belly pain points to a sexually transmitted disease. The nurse chuckles and says it is one of the most frequent complaints they have in this clinic.

I. What is the association between belly pain and STDs?

2. What causes the belly pain, exactly?

3. Is the belly pain a serious sign?

4. What diagnostic tests are called for? Treatment?

5. You look further down on the questionnaire and see there is a question about the client's recent sex partners. She answered that she has had relations only with her boyfriend during the past 12 months. Does the boyfriend need information or treatment? Explain.

Case 6.2

You met your wife Becky at the local hospital where you are a nurse and she works in medical records. You were married 18 months ago and now she discovers she is pregnant. You are both extremely happy and you go together to the obstetrician for her first prenatal visit. Everything looks good; she is seven weeks into a normal pregnancy. About 10 days later you come home from work and she is waiting for you in the kitchen, looking very upset. She tells you that the obstetrician's office called this afternoon with the news that she has tested positive for gonorrhea.

She says it must mean that you have been unfaithful to her, since she knows for sure that she has had no relations with anyone else since you were married. She is so distraught that she will not listen to what you have to say. She packs a suitcase and drives to her mother's house across town.

You believe Becky without question. At the same time you know that you have also been faithful.

1. How is this infection possible if both of you have been monogamous for at least two years?

2. What about the immediate problem? Should Becky be treated, even though she is pregnant? Discuss.

3. Is penicillin the best treatment for gonorrhea? Why or why not?

4. Once these facts are explained to Becky, she calms down. But you wonder if there is a shadow of doubt about your fidelity in her mind. You wonder how other couples fare—particularly those who may have less trust in one another, or don't know how to access information about STDs. Speculate about why Becky's physician did not explain all the possibilities.

Case 6.3

You are working the night shift on The Answer-Line, a telephone service provided by a health management organization for its enrollees. You receive a call from a 26-year-old woman. She has been experiencing painful urination for the past 24 hours. This is the first time she has had this condition and she describes herself as otherwise healthy.

1. What are some of the first questions you should ask?

2. The patient reports a fever of 101.5°F and a nearly constant urge to urinate, though she often voids little or no urine. What is your preliminary diagnosis?

3. There is a certain symptom that she has not mentioned. What is it and why is it important that you ask her about it?

4. What is the most likely causative organism for this condition?

5. What is the route of transmission of this organism?

6. What are some other causative organisms for this condition?

Case 6.4

Your roommate Jane complains to you that she has had intense vaginal itching for the past day and a half. (She often asks you medical questions because you are nearly finished with your nursing degree.) She says the same symptoms crop up from time to time and that she buys over-the-counter antifungal creams that seem to take care of them after a few days. "I guess they are yeast infections," she says. "But I don't know where I keep getting them from." She adds that she hasn't had sex with anyone since her junior prom, three years ago. When you ask her how frequent the episodes are, she says about once every six weeks or so.

1. What is the causative organism of vaginal yeast infections? Where is Jane "getting them from"?

2. What conditions could **predispose** a woman to such frequent yeast infections?

3. Should Jane continue to self-medicate for her yeast infections or should she see a doctor? Please explain.

4. Are there any possible serious consequences of vaginal yeast infections?

Case 6.5

A pregnant woman arrives at your practice because she has noticed a **copious** vaginal discharge and is worried that it may indicate problems with her pregnancy. After a pelvic examination, the physician says there is a whitish, smooth coating on the walls of the vagina. Microscopic examination of vaginal fluids reveals the presence of "clue cells." The physician, a rather gruff type, writes "bacterial vaginitis" on the chart, prescribes an antibiotic, and moves on. The patient turns to you for answers.

1. What usually causes bacterial vaginitis (often called BV)?

2. What are clue cells? What bacterium is associated with clue cells?

3. Is BV a dangerous condition? Explain.

4. Is BV a sexually transmitted disease? Elaborate.

5. What other diseases are in the **differential diagnosis** of a woman with copious vaginal discharge?

6. Is BV treatable? If so, with what? What is the likely outcome of treatment?

Case 6.6

Your best friend, Jack (a 30-year-old investment banker), has had a steady girlfriend for the past six months. He has avoided having sex with her because she told him she has genital herpes. You remember the day that she told Jack about it; he came over to your house very upset and the two of you talked for hours about what that meant for Jack. He thought about breaking up with her because he couldn't see how they could have a long-term, intimate relationship. But finally he decided that he did love her and they would figure something out.

Now she wants to take the relationship to the next level, a level that includes sexual relations. She told Jack that they would do this only if she were lesion-free and that if he wore a condom, they would be fine.

Jack is skeptical. He comes to you for advice.

1. Are they safe if she does not have lesions at the time of their intercourse? Why or why not?

2. Whether Jack's girlfriend has lesions or not, if he uses a condom he will be protected, right?

3. If Jack were the one with herpes and his girlfriend was uninfected, would his use of a condom completely protect her? Explain.

4. What would you say is the safest way for Jack and his girlfriend to have intercourse?

5. What about those new drugs Jack has heard about on TV? Can his girlfriend take those and cure herself? Or at least avoid infecting him? Give some detail.

Case 6.7 Challenge

Your mother told you a story about a 14-year-old patient that she saw in the early 1980s when she was a nurse in a gynecologist's office.

Your mother's first contact with the young girl was after she vomited in the waiting room. She told your mother that she started feeling ill the night before. She had been having unusually heavy menstrual bleeding and reported having a fever earlier that morning. The young patient complained of chills and had a diffuse rash on her arms and legs.

A physician arrived on the scene and he and your mother helped her back to an examining room. Your mother checked her temperature and her blood pressure while the doctor asked her some questions. His first question was whether her neck was stiff or painful. She answered no, but the doctor ordered a **lumbar puncture** anyway. The patient was starting to look dizzy and her blood pressure was low: 90/70. The doctor asked her if she had ever had sexual intercourse and the patient answered that she had not. When the patient's mother came in from parking the car the doctor asked if her immunizations were up to date. The mother confirmed that they were.

Your mom added that the other peculiar thing about this patient was that several days after she was admitted to the hospital, the skin on the palms of her hands began to slough off.

1. What kind of infectious diseases come to mind when a widespread rash is seen as the primary complaint? (Hint: Why had the doctor asked about her sexual history? Why did he ask about her immunizations?)

2. Her rash was diffuse, with well-separated bumps that were **maculopapular.** Was it likely to be chicken pox? Why or why not?

3. The cerebrospinal fluid obtained from the lumbar puncture was clear—no evidence of bacteria. Another infection was ruled out. Which one?

4. The doctor then asked the patient about her menstrual history and practices. She began menstruating at the age of 12 and reported that her last period began four days ago. She reported that she mainly uses tampons during her period. What infection do you think the doctor had in mind in asking about menstruation? What do you know about the infection in question?

5. Your mother says that if you see a patient with these symptoms once you start your practice as a physician's assistant, it is less likely to be the same infection. Why?

Case 6.8 Challenge

A sophomore named Michelle **presents** at the college health service (where you are observing for the day) with fever (105°F), **malaise,** headache, and pain in her genital region that is severe upon urination. The physician examines her genital region and finds four blisterlike lesions on the outer **labia,** each 2–3 mm in size. The lesions are filled with clear fluid; there is no sign of bleeding from them. While you are taking her history, Michelle reports that she has had two successive abnormal Pap smears in the past year. After the first abnormal Pap, she was treated with an antiprotozoal drug. She reports taking the full course of the antibiotic.

1. What questions about the patient's behavior should the physician ask during the history?

2. What is your **presumptive diagnosis** based on the facts presented? What other conditions might be in the **differential diagnosis?**

3. Why do you suppose the patient was treated with an antiprotozoal drug after her first abnormal Pap?

4. Of what importance is the patient's history of abnormal Pap smears?

5. What tests should be ordered to confirm the presumptive diagnosis?

The Usual Suspects

Common microorganisms causing infections in the genitourinary tract[1,2]

Bacteria

Gram-positive

Staphylococcus aureus

Staphylococcus saprophyticus

Streptococcus pyogenes

Gram-negative

Chlamydia trachomatis

Escherichia coli

Gardnerella vaginalis

Haemophilus ducreyi

Leptospira interrogans

Neisseria gonorrhoeae

Treponema pallidum

Viruses

Herpes simplex virus

Human papillomavirus

Fungi

Candida albicans

Protozoa

Trichomonas vaginalis

[1]Not all of the infections appear in this chapter.
[2]Not an exhaustive list.

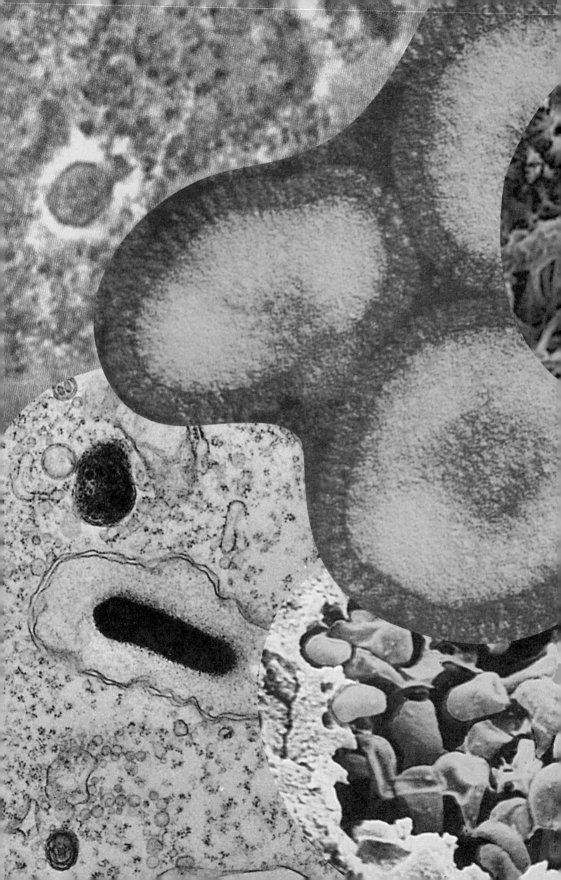

Answer Section

Answers and Explanations for Case 1.1

1. Why would a fertility specialist recommend the rubella vaccine? Why does he suggest a waiting period after vaccination and before conceiving?

Because the rubella virus is particularly dangerous for fetuses. Any person who has no **active immunity** to the virus (because she never had the infection and has not been immunized) runs a small risk of encountering the rubella virus and becoming infected. If a woman becomes infected with rubella while pregnant, her fetus is likely to be infected as well. Although rubella is not a serious disease for children or adults, fetuses may be spontaneously aborted if infected early in gestation. If infection occurs later, deafness, blindness, and damage to nerves and organs may occur. This is called congenital rubella syndrome.

The physician in this case wants to be sure Kate has immunity before conceiving. Because she is seeking medical intervention to conceive, her rubella status can be checked in advance. In fact, all women of childbearing age should be sure their rubella vaccine is up to date. If they are unsure of their immunization history, antibody titers can be checked.

The rubella vaccine consists of live attenuated virus that is safe for administration to children and adults, but is not recommended for pregnant women, as it is more difficult to predict its effects on the fetus. Women have been immunized during pregnancy with no ill effects on their babies, but it is still better to be cautious.

2. When do most children in the United States receive their rubella immunization?

As part of the **MMR** (measles, mumps, rubella) injection, sometime after the age of 12 months, and again at 4–6 years.

3. Kate suggests that she had rubella in second grade, but the disease she described doesn't sound like rubella to you. Why not?

Childhood rubella is almost always very mild. In fact, up to 50% of rubella infections may be **subclinical.** Those who do have symptoms generally present with a very mild rash, beginning on the face and traveling downward. A moderate fever (above 99°F) often accompanies the rash, and there may be some joint pain.

What Kate describes as happening to her in second grade sounds more like measles, not German measles (rubella), although it is difficult to tell so long after the episode. Measles is a much more serious disease than rubella, resulting in a lengthy illness characterized by generalized swelling of lymph nodes, a bright red rash, and a high rate of complications including diarrhea, **otitis media,** pneumonia, and encephalitis.

4. Kate says the doctor gave her the option of having her blood checked for antibodies to the virus, to test her immune status. Would this test be checking for IgM or IgG? Explain your answer.

IgG. IgM antibodies are indicative of new infections.

5. If a physician was checking for a current rubella infection and only had available a test for IgG, how could he or she be certain the infection was a new one?

By testing **acute and convalescent sera.** The presence of IgG to rubella in an ill person's serum does not necessarily mean the person is currently experiencing infection with the virus. The antibody may simply indicate active immunity acquired at some earlier date. But if the level of rubella-specific IgG rises significantly over the course of 10–14 days, it can be assumed that the patient is currently mounting a specific immune response.

Answers and Explanations for Case 1.2

1. On the basis of Meg's **oral history,** what is the most likely diagnosis? What would the causative microorganism look like in a Gram stain?

Contact dermatitis is unlikely if Meg has not recently been exposed to new fabrics. Her history of being in the woods, as well as the radiating appearance of the rash, strongly suggests Lyme disease, caused by *Borrelia burgdorferi*. The rash is often circular in shape and, over time, spreads outward from the initial site. However, it is not always present and is often less dramatic than the bull's-eye form seen in textbooks.

Borrelia is a gram-negative spirochete. It would look like pink corkscrews in a Gram stain.

2. How did she most likely acquire her infection?

Lyme disease is acquired through the bite of a tick carrying *Borrelia*, a bacterium with a complex life cycle. The ticks (genus *Ixodes*) serve as intermediate hosts of the bacterium. The normal hosts of the ticks are deer and small animals and birds. *Ixodes* ticks can be very small. One mother reported thinking her daughter, who was infected by a bite on the face, had simply acquired a new freckle! Although Meg does not remember being bitten, the most common place to find Lyme disease lesions is on the buttocks and the back of the thighs. This is because it takes about 48 hours for the tick to transmit the bacterium to its human host, and people are more likely to find and remove a tick from more visible parts of the body.

3. Would the diagnosis be any different if Meg had attended camp in Arizona? Explain.

Cases of Lyme disease are much more common in the northeastern United States than in the Southwest. If Meg had spent the month of August in Arizona, it would be less likely that her rash was due to *Borrelia* infection. In that case, more sophisticated laboratory diagnostic techniques would probably be performed to confirm the diagnosis.

4. Why does the doctor ask Meg if her joints hurt?

The doctor knows that if Lyme disease goes untreated, it can later cause neurological symptoms and arthritis-like pain in the joints. It is a good sign that Meg still has early-stage symptoms (the rash) and hasn't experienced later symptoms.

5. How is this infection treated?

Early-stage Lyme disease can be treated and cured with antibiotics. If it has progressed to later stages, antibiotics are not effective.

6. Meg's mom, upon hearing the **presumptive diagnosis,** declares that Meg will not return to that camp, which she loves and had planned to attend next summer. The doctor suggests that Meg need only take some precautions. How can she protect herself from getting this infection again?

Meg can wear protective clothing, such as lightweight long pants, when she is in the woods, and cover unprotected skin with an insect repellent. Alternatively, a new **recombinant** vaccine for Lyme disease is recommended for people over the age of 14 who have prolonged exposure to high-risk areas.

Answers and Explanations for Case 1.3

1. What disease do you suspect? Explain why.

You most likely have a fungal infection of the fingernail. This condition is called *onychomycosis* and can be caused by a wide variety of dermatophytes. Other microorganisms find it very difficult to thrive on **keratinized** tissue. You need not be embarrassed; this is an extremely common infectious disease. Artificial nails can **predispose** you to fungal infections if pockets are left between the artificial and real nail. These pockets trap moisture (and organic substances) and provide favorable environments for fungal growth.

2. What would you suggest be done for a more definitive diagnosis?

Many conditions can cause nails (both toenails and fingernails) to have an altered appearance; fungal infection is only one of them. Definitive diagnosis requires verifying that fungi are present in the affected area. This can be done through microscopy of a nail scraping, potassium hydroxide screening, culture, or **histological** examination of a biopsy sample.

3. Can you treat this yourself with an over-the-counter drug, or do you need to see a physician?

It is best to visit a doctor and get a diagnosis (and prescription, if necessary). Abnormalities of the fingernails could indicate other types of metabolic conditions or **chronic** diseases.

4. You see cures for this condition mentioned on TV and on the Internet—do you think they work?

Fungal nail infections can take many forms. The mildest forms can often be cured with topical preparations of antifungal drugs such as itraconazole or terbinafine. Research has also shown that some natural substances, such as tea tree oil, can be effective. Over-the-counter medications containing one of these ingredients may or may not be effective. Additionally, as discussed in the answer to question 3, it is best to let a physician see the affected area to make a definitive diagnosis.

The most important thing to remember is that treatment must be continued for at least 3 months. Many researchers suggest that 12 months is required for complete destruction of **dormant** fungal forms in the nail tissue.

Complete cures can be elusive, even with continuous long-term treatment. Oral antifungal drugs are used in more serious cases but do not always result in cures.

5. What other conditions are caused by **dermatophytes** (*Microsporum, Trichophyton,* and *Epidermophyton*)? What is special about them that makes them capable of thriving in their anatomical **niche** on their hosts?

These fungi cause surface **mycoses** of all types, such as athlete's foot (called *tinea pedis*), jock itch (*tinea cruris*), and ringworm. Surface mycoses are restricted to the body's surfaces, and are difficult to resolve. They have the ability to digest keratin, the protein packed into cells making up the hair, nails, and upper layers of the skin.

Answers and Explanations for Case 1.4

1. What is the most likely diagnosis, and what is the **etiology**?

Bacterial conjunctivitis. The condition is commonly called pinkeye, for obvious reasons. Conjunctivitis can be bacterial or viral in origin.

The most common bacteria causing it are in the genera *Haemophilus*, *Streptococcus*, and *Staphylococcus*. Adeno- and echoviruses are responsible for most viral conjunctivitis cases.

2. What sign leads you to believe that the infection is bacterial in origin?

Thick yellow or white discharge usually suggests bacterial infection. Viral conjunctivitis infections most often cause a clear watery discharge.

3. What is the treatment for this condition? Elaborate. Is the condition **communicable**?

Bacterial conjunctivitis can be treated with antibiotic eye drops or creams. Keisha should be kept home from school until she has been treated for 24 hours, as conjunctivitis is highly contagious.

In any given child, an infection in the upper respiratory route can spread to the eye via the nasolacrimal duct, causing conjunctivitis. But once a child has the symptoms in her eye it can be directly transmitted to another person's eye.

4. What are some of the eye's natural defenses that help to prevent infections?

The conjunctiva itself is composed of **epithelium,** which is somewhat protective. Flushing of the eyes with tears is a very effective defense. In addition, tears contain **lysozyme,** an enzyme that can break down peptidoglycan.

5. Are there steps the teacher should take to prevent the spread of this infection in the classroom? If so, discuss them.

Conjunctivitis is spread through hand to eye contact. Nasal secretions can carry the microorganism and often find their way to the eye. Items that are shared by children should be disinfected, if possible. Where possible, disposable items (such as paper towels) should replace shared items. Children (and adults in the classroom) should be encouraged to wash their hands frequently.

The teacher should also send a note home to all the students' parents to watch for signs of pinkeye in their children and in **household contacts.**

Answers and Explanations for Case 1.5

1. What condition did this patient have? What features suggest that it is not *Clostridium perfringens* gangrene?

Necrotizing fasciitis (NF) caused by group A streptococcus (*Streptococcus pyogenes*). This is the condition that the media call "flesh-eating disease." It is one of the invasive diseases that can be caused by group A strep, the bacterium most commonly associated with causing sore throats.

Another invasive manifestation of group A strep infection is streptococcal toxic shock syndrome (STSS), a bloodstream infection with a high fatality rate.

The fact that the original wound was not a deep anaerobic wound is evidence against *C. perfringens.* Also, gram-positive cocci were isolated from the wound. *C. perfringens* is a rod-shaped spore former (which would be visible on the Gram stain).

2. Why was amputation the best solution for the infection in this case?

Necrotizing fasciitis is a fast-spreading condition that is fatal in 25% of the cases, even with aggressive antibiotic therapy. The bacteria are extremely difficult to eliminate from the tissue, and will very rapidly destroy (necrotize) skin and connective tissue. Surgical debridement (removal of affected tissue) increases the chances of successful antibiotic therapy, but complete removal of damaged tissue plus a small border of unaffected tissue is often necessary.

3. How is the bacterium transmitted?

The bacterium is transmitted through person-to-person or **fomite** contact. You cannot get this disease from eating contaminated foods. (An email hoax in the summer of 2000 suggested that Costa Rican bananas can cause NF!) A substantial portion of the healthy population carries group A strep, either in their nasopharynx or on their skin. Even the most **virulent** of these strains cannot cause necrotizing fasciitis in intact skin. There must be trauma or breaks in the skin (surgery, sports injury, shaving wound) for the bacterium to enter underlying tissue and cause disease. Although not foolproof, good hygiene is the best prevention for this condition.

4. It seems like we've heard a lot more about this condition in the past few years. Is this just media hype or are more cases occurring? Explain.

There certainly has been a lot of media attention to this sensational disease. But it appears that the increase in cases is real. Although most group A streptococcus infections continue to be uncomplicated, certain strains of the bacterium have increased in virulence and are able to cause aggressive disease.

Answers and Explanations for Case 1.6—In the News

1. What was this resurgent infection?

Measles.

2. What are some possible reasons for the epidemic in 1989–1991?

The incidence of measles had been very low in the 25 years since the introduction of the vaccine in the United States. When the incidence suddenly increased, people looked to the vaccine for clues. Either the vaccine was no longer effective (Can you think of explanations for that?)* or not enough people were receiving the vaccine.

In this case it appeared that low vaccination coverage was responsible for the epidemic. In some cities, fewer than 50% of

---

*One way vaccines can lose their effectiveness is if the naturally circulating pathogens change their antigens so that they no longer match the ones in the vaccine preparation. Then the immunized person's antibodies won't bind to the new or altered pathogen. Second, technical problems with the vaccine should be considered. Has something changed in the manufacture of the vaccine? Are the problems confined to one lot of the vaccine? If so, had the lot expired? Was it stored (refrigerated) properly?

2-year-olds had received their **MMR** (measles, mumps, and rubella) immunization. This left a large susceptible population and facilitated the spread of the virus. These children were more likely to be vaccinated once they started school, which may explain the shift in incidence of the disease away from school-age children to the pre-school-age group.

Another interesting factor was that **passive immunity** to the virus was lower among babies of mothers whose own immunity to the virus was acquired from vaccination rather than from active infection. Because the vaccine was introduced in 1963, women born before that time were likely to have had active measles. Mothers who had been vaccinated against measles when they were children were spared the disease, but may have had a slightly less effective antibody response to pass on to their own fetuses. These babies needed to receive the measles vaccine themselves in order to be protected from the disease.

This epidemic illustrates that it is easy to become complacent about vaccinations when the incidence of a disease has been very low for a long time. But immunization continues to be absolutely critical.

3. What is **herd immunity**? Discuss it in relation to this outbreak.

Herd immunity refers to an entire population being protected from an infectious disease even though not everyone is immune. If enough members of a population are immune, due to either previous infection or immunization, then the microorganism cannot sustain itself via transmission among the population. So even those who have had no exposure to the infection or the vaccine are not likely to get sick. The relative proportion of a population that has to be immune in order to protect "the herd" depends on the infectious agent.

With rates of measles vaccine coverage as low as 50%, a large portion of the population was susceptible. These people can be infected, and can pass the virus on to others.

4. What is the schedule for vaccination for this infection in the United States?

The measles vaccine (MMR) is to be administered at the age of 12–15 months and then again between the ages of 4 and 6 years (or definitely by the age of 11 to 12 years).

5. Are serious **sequelae** associated with this infection? If so, what are they?

Measles is not a mild disease. Even in the absence of complications, patients are quite ill during a measles episode. The most common complications are rather mild, such as diarrhea or otitis media. Pneumonia occurs in approximately 7% of measles cases and can be serious. One out of 1000 measles cases results in encephalitis. The symptoms associated with encephalitis may be headache, fever, vomiting, stiff neck, drowsiness, convulsions, and coma. This complication can be fatal in a small percentage of the cases.

Answers and Explanations for Case 1.7—Challenge

1. What are the lesions diagnostic of? Explain how you decided.

Shingles (herpes zoster). A group of small **serous** lesions confined to one side of the body (i.e., the group does not cross the midline) is usually a late-stage manifestation of infection with the varicella-zoster virus, a member of the herpesvirus family.

2. Although this particular condition is somewhat unusual in babies, the lesions indicate that the child must have experienced a common childhood illness earlier. Which one?

Chicken pox.*

3. The mother says that, to her knowledge, the baby has not had this common childhood illness, but that his 3-year-old sister had it

---

*Have you ever wondered why this disease is named after chickens? Chickens are certainly not the **reservoir** for the virus. Humans are. Most scholars think the name came from the Old English word *giccan,* which means "to itch." Giccan, chicken . . .

Answers and Explanations for Case 1.8—Challenge

1. Your diagnosis? Why was his throat checked?

Fifth disease. This disease acquired its name 100 years ago when it was discovered. At the time of its discovery there were four other common causes of childhood rashes: Filatow-Dukes' disease (previously known as roseola), scarlet fever, measles, and rubella. Scientists must have been particularly uninspired at the time and came up with this name. (It is also called erythema infectiosum, but not very often.)

As for the **differential diagnosis** in this case, the first hallmark is the slapped-cheek appearance of the rash. Because his vaccinations (including the MMR) are up to date, measles and rubella are unlikely. Filatow-Dukes' disease is usually accompanied by a high fever. It is unlikely that he has chicken pox, because he has already had it and the symptoms don't match (the type of rash, his apparently feeling well). Scarlet fever is also unlikely if he has no history of a sore throat and is not ill currently. (His throat was checked to rule out scarlet fever.)

Fifth disease, most common in late winter and spring, is caused by parvovirus B19. (It is a human parvovirus—you cannot get this disease from a dog or a cat.)

2. Can John go to kindergarten today? Why or why not?

Yes, he can. By the time the rash appears with fifth disease, the patient is no longer contagious (or ill). Of course, the teacher is owed an explanation, but he or she should be familiar with the disease.

3. Is this infection rare? Explain.

Symptomatic cases of fifth disease may not be common, but the infection is. The Centers for Disease Control and Prevention (CDC) estimates that up to 50% of adults have had this infection during their life.

four months ago, when the baby was 2 months old. Explain the link between the girl's illness and the baby's condition.

Shingles can occur only if a person has been infected with the varicella virus. Initial varicella infections almost always cause obvious chicken pox, but in this case it seems to have infected the baby without causing the early symptomatic form of the disease (chicken pox).

4. What factors probably influenced the fact that the baby did not have symptomatic illness when his sister was experiencing it? And what factors led to the eruption of lesions now?

It is likely that when his sister was contagious with chicken pox the baby was partially protected by passive antibody acquired from his mother *in utero.* If the mother was breast-feeding the boy he would be receiving additional passive antibodies through the breast milk. Although at the time it seemed that this protection kept the baby from being infected, it actually served only to prevent symptomatic disease. The baby was infected, and the virus then became **latent** in his spinal ganglia.

By the time the baby was 6 months old, levels of passive antibody acquired *in utero* had declined greatly. The mother also reported she stopped nursing when the baby was 4 months old because she returned to work. The antibodies acquired from the mother had apparently been holding the virus in check, but now that they had dropped below a threshold level the virus was able to manifest itself in its late-stage form as shingles.

5. Is this a dangerous condition? Why or why not?

No. The lesions will heal in a few days. The child will probably not experience recurrences of the lesions because in healthy people with intact immune systems shingles tends to occur only once. Elderly people and those with weaker immune systems can have recurrences of shingles.

4. Are any **sequelae** associated with this infection? If so, name them.

There are almost never sequelae for young children who acquire this infection. When adults get the disease they sometimes suffer short-term joint discomfort, but it doesn't last. There are slight risks to a fetus who is exposed to this virus *in utero.* If a woman becomes infected during her first trimester she has a slightly increased risk of miscarriage. Later in pregnancy, maternal infections can result in fetal anemia.

Answers and Explanations for Case 2.1

1. What type of information is contained in all vaccine brochures? Why should they be read before the vaccines are administered?

In a small percentage of cases, vaccines can cause side effects or, in the case of live vaccines, symptoms of the disease they are meant to prevent. The written information the mother was given describes normal reactions to the vaccines, as well as serious side effects.

Many medical treatments carry risks. The risks are more acceptable when the treatment is given to alleviate disease and/or symptoms. Because vaccines are administered to healthy patients, the medical profession has an even greater responsibility to ensure that patients are informed of the possible problems associated with them.

2. What particular facts are critical for parents to know about the polio vaccine?

The child in this scenario would be given the oral polio vaccine, which contains live (attenuated) poliovirus. It has been documented that the vaccine strain of the virus can (very rarely) cause polio in the person vaccinated or his or her **household contacts,** particularly in those who have weakened immune systems. The Centers for Disease Control and Prevention (CDC) estimates that the rate of occurrence is 1 in 2.4 million vaccinated children. Therefore, if the child to be vaccinated lives in close contact with anyone who is undergoing **chemotherapy,** has natural or acquired immune deficiencies, or is elderly, or if the child has any of these conditions, the use of the killed (injectable) version of the polio vaccine should be considered.

3. What other vaccines besides the MMR and polio are appropriate for a 1-year-old child?

The Hib vaccine (*Haemophilus influenzae* strain b) is standard at 12 months. In addition, the CDC recommends the hepatitis B (HepB) and varicella (chicken pox) vaccines.

4. Two forms of the polio vaccine are available, the live attenuated version (called the OPV, or oral polio vaccine) and the killed or inactivated version (called the IPV, or injectable polio vaccine). Why is the OPV the preferred version for this age?

The live attenuated version (OPV) is preferred for two reasons, despite its higher risk of complications as compared to the IPV. First, its oral format means one less injection for the child. Second, it is at least theoretically more effective than the inactivated vaccine because the live virus will replicate and continue to stimulate the immune response. Third, it is believed that live vaccine versions of the virus, which are **shed** by the vaccinated person from the gastrointestinal tract, contribute to "passive" vaccination of the population, thereby ensuring that **herd immunity** will be high enough to prevent the resurgence of polio infections.

Answers and Explanations for Case 2.2

1. What is the most likely causative organism? Why?

Haemophilus influenzae. It is a **pleomorphic** gram-negative bacillus. Both *Neisseria meningitidis* and *Haemophilus influenzae* grow on chocolate agar, but the two can be distinguished by their Gram staining characteristics. *Neisseria meningitidis* is a diplococcus. (*Streptococcus pneumoniae*—another possible cause of meningitis—is also a diplococcus, but because it is gram-positive it won't be confused with *N. meningitidis*.)

A simple Gram stain can tell you a lot! In this case, antibiotic therapy would be started immediately, on the basis of the initial Gram stain of the original CSF.

2. Why was the child's unvaccinated status helpful in diagnosis?

Because most children receive a series of immunizations with a conjugate *H. influenzae* type b, or Hib, vaccine. The first Hib vaccine was introduced in 1985. Improved vaccines have become available since then and are part of the recommended childhood immunization schedule. Before vaccines were available, *H. influenzae* infections were the most common cause of meningitis in children. Since 1985, the incidence of H. influenzae infections of all types have decreased by at least 95%. However, the fatality rate for the infection is 2–5%, even with appropriate antimicrobial therapy, and a significant percentage of patients experience long-term neurologic **sequelae.** Because many children are inadequately vaccinated, clinicians and diagnostic labs must still consider it in the **differential diagnosis.**

3. What is causing the cloudiness in the CSF?

The cloudiness is caused by the accumulation of bacteria and white blood cells in the CSF. The CSF will have a different complement of white blood cells in it depending on what type of organism is causing the infection. For instance, bacterial meningitis is accompanied most often by increased **PMNs,** whereas increased **lympho-cytes** are seen more often in fungal meningitis.

4. What other types of infections can this causative organism cause in children?

H. influenzae can cause **epiglottitis, otitis media,** and pneumonia. It was named H. influenzae because it was identified in influenza patients. As it turns out, the bacterium was causing a secondary lung infection; the patient was initially afflicted with influenza caused by the influenza virus.

Answers and Explanations for Case 2.3

1. What is the first step in determining if the patient has meningitis?

Lumbar puncture. Cerebrospinal fluid must be examined under a microscope to look for the presence of microorganisms and white blood cells.

2. This test reveals the presence of very large cells that appear to be eukaryotic, surrounded by a very large capsule. What is the probable diagnosis? Name some other eukaryotic organisms that can cause meningitis symptoms.

It is most likely Cryptococcus neoformans—an **opportunistic** fungal pathogen. Other eukaryotic microorganisms sometimes found to cause meningitis include *Coccidioides, Histoplasma,* and *Toxoplasma. Naegleria fowleri* (an amoeba) can also cause a condition called primary amoebic meningoencephalitis (PAM).

3. What groups of people are at risk for this infection?

Immunocompromised people. Eighty-five percent of people infected with this fungus also test positive for human immunodeficiency virus (HIV); other immunocompromised groups may also become infected.

4. How is it acquired?

It is acquired through the respiratory tract. Soil contaminated with infected bird droppings is the usual source. When the soil is disturbed or the dust becomes airborne, it is easy to inhale.

5. What anatomical sites are most often infected with this fungus?

The central nervous system, as seen here, and the lungs. Pulmonary infections are very common.

6. Let's say your initial suspicion (your answer to question 2) was correct. What other diagnostic test should be performed on this patient?

An HIV test. Cryptococcal infection alone is usually enough to warrant an HIV test, and this patient has several other signs that indicate acquired immune deficiency syndrome (AIDS). The lesions on his mouth may signal herpes infection, the symptoms of which are more severe in patients whose T-cell numbers are low. And the dark spots on his face and upper body may be Kaposi's sarcoma, a malignancy associated with AIDS.

Answers and Explanations for Case 2.4

1. What is your suspicion, based on what seem to be nervous system symptoms?

You suspect the child has infant botulism, a paralytic disease caused by **exotoxins** of *Clostridium botulinum,* an anaerobic soil bacterium.

2. If this is indeed the case, do you start treatment here at Kevin's home, or should you transport him to the local hospital?

If you suspect botulism, immediate hospitalization is called for. Also, the doctor must call the state health department immediately for further guidance, as it is charged by the Centers for Disease Control and Prevention (CDC) to oversee these cases. It is imperative that the child receive the correct care immediately. It is also imperative that the state health department be notified so it can launch an investigation of the source of the organism in order to prevent infection of others and to be alert for such infections.

3. What should be administered to Kevin at the earliest opportunity?

Antitoxin to the botulinum toxin, available from the CDC. (Again, state health departments can arrange for this).

4. How do babies acquire this condition?

Through ingestion of *C. botulinum* spores or vegetative cells, which colonize the intestines and then produce toxin. This is different from foodborne botulism, in which the preformed toxin from *C. botulinum* is ingested by the host. Honey has been implicated as a source for the organism and is therefore not to be fed to babies. But up to 80% of cases of infant botulism occur in babies with no history of honey consumption. In these cases, exposure may have occurred through exposure to soil or blowing dust.

5. Although the diagnosis should be confirmed with laboratory tests, the tests should probably not be performed in the hospital lab. Why not?

Because the botulinum toxin is extremely potent in very tiny quantities. Only labs and personnel experienced in isolation of highly dangerous pathogens should handle body materials containing the toxin or bacterium. Many state labs can do this, as can the CDC. Again, the state lab will guide this process.

Answers and Explanations for Case 2.5—In the News

1. What type of organism would you look for in a Gram stain of blood or cerebrospinal fluid in these cases?

A gram-negative diplococcus. The meningococcus is *Neisseria meningitidis.*

2. What is the organism's **portal of entry** to the host?

The respiratory tract.

3. Could you swab the portal of entry (see question 2) to detect the presence of the organism? Why or why not?

No, a throat or nasal swab is not useful. *N. meningitidis* can be normal flora in healthy people; detecting it in the respiratory tract does not indicate that it is the disease-causing organism.

4. What types of symptoms are associated with meningococcal disease?

Bloodstream infections, called meningococcemias, and meningitis. Meningococcemia is characterized by a **petechial** rash and low blood pressure, and possible long-term **sequelae** such as **peripheral** limb paralysis. Meningitis caused by *N. meningitidis* also produces the rash in addition to classical meningitis symptoms: stiff neck, headache, confusion, and high fever. The fatality rate is high and, even with aggressive medical treatment, can be as great as 10%.

5. A total of 900 students attend the affected middle school. What measures should have been taken to protect the remaining 895 students from acquiring meningococcal disease?

Prophylactic antibiotic treatment of school and **household contacts** is always recommended in outbreaks of meningococcal disease. Rifampin is often the drug of choice. It should be administered to contacts within 24 hours of identification of a case.

Answers and Explanations for Case 2.6—In the News

1. How is West Nile virus transmitted?

West Nile virus is a mosquito-borne disease.

2. The virus had another vertebrate host besides humans when it showed up in New York. What was it?

Birds. West Nile virus can infect a wide range of vertebrates, but in the United States, unlike in its normal geographic homes, it was found to be killing birds. In its usual environment, the virus infects but seldom kills birds. This suggests a change in the **virulence** of the virus.

3. Can you list some possible mechanisms for how the virus was introduced into the United States?

No one knows how the virus was actually introduced, but the following possibilities were considered during this epidemic. An infected person could have traveled to the United States and infected mosquitoes, which in turn infected birds, and then other mosquitoes. Alternatively, infected mosquitoes or migrating birds themselves could have traveled to the United States. A third possibility is that an infected bird was imported as part of the commercial bird trade.

4. Most infections result in no noticeable symptoms. Some of those infected may develop a skin rash. A fraction of people infected develop life-threatening encephalitis. What is encephalitis and who do you suppose is most likely to experience this symptom?

Encephalitis is an inflammation of the brain. People over the age of 50 and people with weakened immune systems are most likely to experience this severe form of the infection. Very young children may also be at higher risk for serious disease.

5. A sudden increase in a particular disease within a population of humans is called an **epidemic.** What is a large outbreak among animals called?

In the weeks preceding the park closing, 40 birds infected with West Nile virus had been found in eight counties surrounding New York City. Health officials considered this an epizootic.

6. If you lived near Central Park and wanted to go jogging there, what would be the best time of day to avoid the park to minimize your chances of being infected with West Nile virus?

Mosquitoes are most active at dusk, when they begin feeding. They remain active until dawn. It may be better to jog in the morning or at midday.

Answers and Explanations for Case 2.7—In the News

1. What is the name for a transmissible agent that contains only protein and has no genetic material?

Prion.

2. What is the formal name for mad cow disease? Explain the name.

Bovine spongiform encephalopathy. Bovine—in cattle; spongiform—resembling a sponge, referring to brain tissue that is pitted with holes by the infection; encephalopathy—disease of the brain.

3. The human form of the disease is called something else. What is it?

Creutzfeldt-Jakob disease (also spelled Creutzfeld-Jacob).

4. Scientists suspect that the humans infected during this outbreak acquired the disease from eating meat from diseased animals. Even when meat is well cooked, it transmits the infection. What does this say about the infectious agent?

The agent is not destroyed by heat. Nor is it destroyed by most other forms of sterilization. One scientist claims that he has exposed the agent to 600°F for a period of 15 years without destroying it.

5. These cases in Britain were not the first cases of the disease; it occurs at a low constant rate in other countries, including the United States. Although some of these sporadic cases can be traced to transplants of infected tissues, such as corneas or brain tissues, most are idiopathic. What does *idiopathic* mean?

Idiopathic means "of unknown origin or causation."

6. Livestock control measures have been in place in Britain for several years now. Can we expect more human cases with links to the British cattle epidemic, or is it behind us? Defend your answer.

Because infected cows continued to turn up in the late 1990s, and because the incubation period for the disease is approximately 10 years, we will no doubt continue to see human cases of the disease for years to come.

Answers and Explanations for Case 2.8—Challenge

1. The doctor speaks with you in the hallway. He tells you that mental illness is very common among the homeless population. Dwight needs a thorough psychiatric evaluation. The doctor is fairly sure that some of Dwight's neural symptoms are caused by a sexually transmitted infection. Which one? Caused by which microorganism?

A blood test for syphilis, caused by *Treponema pallidum,* is in order. In 5–10% of people, the tertiary stage of this disease can involve the nervous system, resulting in symptoms resembling severe mental illness. Syphilis is more common in the homeless population than in the general population.

2. If the blood test comes back positive, does it mean that Dwight can transmit the disease to others? Explain.

No. Syphilis infections are transmissible only when **mucocutaneous** lesions are present. These occur in the primary (chancre) state and in the secondary phase. This patient is in the late tertiary stage.

3. Should Dwight be treated with antibiotics to remedy his neural symptoms? Why or why not?

Most physicians believe antibiotic treatment in the tertiary phase is ineffective. *T. pallidum* cannot be isolated from patients in this stage and the damage is probably induced by the body's own chronic inflammatory response.

4. The patient's blood test came back positive. For what other infectious disease should he now be tested?

HIV. Patients with syphilis are often coinfected with this virus.

5. The doctor tells you to expect to see more of these cases in the future. But a coworker, who graduated from nursing school 10 years ago, tells you that this disease (especially its later forms) is relatively rare and is decreasing in incidence. Who is right?

The physician may be right. Public health experts have warned health providers to be on the lookout for tertiary syphilis because many people with AIDS are staying alive longer and their syphilis infections may reach the tertiary stage.

Not all syphilis cases progress to the tertiary stage. In the early 1900s, 33–50% of infected people eventually experienced the destructive tertiary form of syphilis. The advent of antibiotic therapy sharply reduced that number. Interestingly, even when antibiotics are not used to eliminate the bacterium in the early stages, a lower percentage of people today seem to progress to the tertiary stage compared to a few decades ago. This is seen as evidence that the bacterium may be evolving to a lower virulence.

Disease progression in the HIV-positive person is a different story. Some evidence indicates that even those who are treated adequately in the early stages of syphilis have an increased chance of tertiary stage symptoms compared to HIV-negative patients. (Can you speculate on why that may be?)

Answers and Explanations for Case 3.1

1. Which cardiovascular infectious condition is this?

Subacute bacterial endocarditis (SBE).

2. What is the most likely causative organism and the route of transmission?

Streptococci that are normal flora in the oral cavity are the most common cause of SBE. They are transmitted to the heart when they are introduced into the bloodstream. This can happen during routine dental procedures or during surgery involving the upper respiratory system.

3. What's the connection, if any, with rheumatic fever?

The causative bacteria attach to heart valves as they pass through the heart. However, the bacteria don't easily attach to ordinary heart valves. Damaged or misshapen valves display surfaces to which these bacteria can strongly attach. An earlier episode of rheumatic fever is one of the ways heart valves become susceptible to colonization with this type of streptococci.

4. Why did you look at his fingernails?

Petechial hemorrhages in the fingers and toes, or even the eyes, are a sign of SBE. They are caused by the presence of bacteria in the capillaries, which leads to localized inflammation and leakage of fluids from the vessels.

5. What type of culture would a physician most likely order, and why?

Cultures on blood agar. These bacteria are **alpha-hemolytic,** meaning a green or brown halo forms around the colonies when they are growing on blood.

6. What is the treatment? Is there a way to prevent the condition?

Patients diagnosed with SBE should be hospitalized and given intravenous antibiotics. People who know they have valve abnormalities should receive **prophylactic** antibiotics before any procedure that may introduce bacteria into their bloodstream. This is why many people are prescribed a short course of antibiotics before seeing the dentist.

Answers and Explanations for Case 3.2

1. What is your diagnosis?

Dengue fever, caused by an arbovirus (arthropod-borne virus). This diagnosis seems very likely because of the combination of symptoms and geography. High fever, severe pain, and rash are characteristic of dengue. And the victim comes from El Salvador, where dengue is **endemic.** El Salvador also experienced an **epidemic** of dengue in the year 2000.

2. What connection does this disease have to broken bones?

The disease is often called "breakbone fever" because of the severe pain it causes in deep tissues. Bones do not actually break.

3. This is a vector-borne disease. What is the name of the most common vector?

Aedes aegypti, a type of mosquito, is the most common vector for this virus, although other mosquitoes can transmit it as well.

4. What other infection is transmitted by the same vector?

Yellow fever, caused by another arbovirus.

5. The next night when you arrived at work the little girl seemed to be doing better. The rash had subsided and her fever had lowered. But on the third night you arrived to find that she had been transferred to intensive care after hemorrhaging internally. Is this still consistent with your original diagnosis? Explain.

Most dengue fever cases resolve themselves without serious complications. But in a form of dengue called dengue hemorrhagic fever (DHF), an apparently recovering patient takes a sudden turn for the worse and begins to bleed internally. A significant percentage of these patients die. The hemorrhagic form of dengue was seen quite frequently in El Salvador during 2000.

6. This all sounds very bad, but you're somewhat comforted by the fact that this disease is not found in the United States. Right?

Right—for now! But the presence of the vector is the most important factor leading to introduction of an arbovirus in a geographical area. *Aedes aegypti* exists in some southern U.S. states. And *Aedes albopictus* mosquitoes (which are highly capable of carrying this virus) are in over half of the states in the United States Surveillance is ongoing to monitor the incidence of dengue fever (and yellow fever) in the United States.

Answers and Explanations for Case 3.3

1. On the basis of the limited information above, the physician has obviously made a diagnosis. What is it? What does it look like when Gram stained?

Tularemia caused by *Francisella tularensis,* a gram-negative coccobacillus.

2. What is the most likely reservoir for the causative organism in this case?

The rabbit. Tularemia is usually a **zoonosis,** most often found in wild animals such as squirrels, rabbits, and deer. It can be transmitted when infected animal materials are introduced into deeper tissues of a human. The eye could also be the **portal of entry.** Rarely, insect vectors can also transmit it by biting or stinging.

3. Why draw blood twice?

To obtain **acute and convalescent sera** to check for levels of anti-*Francisella* antibodies. If the titer rises in the second sera, it is indicative of current infection. Although tularemia is not a **notifiable disease,** it is helpful to definitively diagnose the condition, as it frequently recurs even after antibiotic therapy has cleared the initial signs of it. Having a definitive diagnosis will help the physician manage the infection if it recurs.

4. Why not culture the wound to look for the bacterium?

Francisella does not grow on any common culture media. Furthermore, it is extremely dangerous to handle in a laboratory setting. For these reasons, most diagnostic laboratories are not equipped to culture the bacterium.

5. What are some other common infections that humans acquire from animals? (These are also known as **zoonoses.**)

The major zoonoses in the industrialized world have long been tularemia, brucellosis, and rabies. Anthrax is also associated with animal transmission to humans. The plague (*Yersinia pestis*) can also be transmitted from small wild animals (such as prairie dogs) to domestic pets and then to their human owners via fleas.

Health care providers should be on the lookout for other, emerging zoonotic infections. Hantavirus (associated with rodents), *Salmonella* infections (in chickens and reptiles), and the West Nile virus (in birds) are examples.

Answers and Explanations for Case 3.4

1. What is the name of the condition you suspect?

Malaria is the most likely condition. (It is not the name of the causative organism, however.)

2. What should be your first question about the patient's history?

Had he traveled internationally just prior to the onset of this illness two months ago? Malaria is no longer present in North America. Although a few cases appear every year in the United States, the origins of the infections can usually be traced to other countries. For instance, immigrants from other countries can be ill with malaria they acquired while living in their native countries. The biggest risk to those living in the United States is travel. Most countries in the Southern Hemisphere have malaria, and travelers should take steps to protect themselves.

3. What is the most likely causative organism (genus and species)? Support your answer.

Plasmodium malariae. This protozoan cycles into host red blood cells and ruptures them approximately every 72 hours. Other species of *Plasmodium* have different lengths of this erythrocytic cycle.

4. Is this pathogen eukaryotic or prokaryotic?

It is a protozoan, a single-celled eukaryotic microorganism.

5. Which is the most dangerous of the species that can cause this disease? Give some details.

Plasmodium falciparum. This species is often fatal. It can cause damage in many different organs because it has the ability to **occlude** capillaries. The most dangerous manifestation of this is called cerebral malaria, a condition where infected red blood cells stick to capillary walls and eventually close the vessel off. Areas of the brain (or other organ) served by those blood vessels then die. Any species of *Plasmodium* can cause this, but it is far more common with *P. falciparum.*

6. What are the two main places in the human body that are exploited by the causative organism in this disease?

The liver and red blood cells. The life cycle of the protozoan is very complex; refer to your textbook for a review.

7. Can this individual transmit this infection to others? Why or why not?

This disease is not transmissible via direct contact. Vector transmission by the *Anopheles* mosquito is by far the most common way to acquire malaria. In rare instances blood transfusions and the use of dirty needles have transmitted the infection as well.

Answers and Explanations for Case 3.5

1. In your first week you saw several children whose major symptoms were high fever, lots of sweating, and **prostration.** They all turned out to have the same infectious condition, one that you continued to see throughout your stay in South Africa. Up to half of the sick children did not survive this illness. What is it?

Malaria. It is estimated that one in twenty children in the southern part of the African continent die of malaria before the age of 5. At any given time 10–20% of them have active infections.

2. In this setting, what is the best prevention for this disease?

The use of bed nets. The only way to avoid getting malaria is to avoid being bitten by the *Anopheles* mosquito. Mosquitoes are more active after dusk, and sleeping under bed nets has been found to reduce the likelihood of infection. Bed nets impregnated with insecticides are even better, but are too expensive for many families.

3. In your third month you saw a 2-year-old boy with an angry-looking rash. He was very ill with a high fever, and eventually died. His death surprised you because you thought this disease had been conquered long ago. (In the United States it is seen only occasionally, because children are vaccinated for it.) Over the course of your two-year stay you saw these symptoms in children perhaps a dozen times. Several of the children died. What is the disease?

Measles. Vaccination rates for this disease are relatively low in developing countries; thousands of children are killed by measles every year.

4. Name at least two of the most common infectious conditions you should look for in adult clients in this setting.

Human immunodeficiency virus (HIV) is highly prevalent in South Africa. Among pregnant women, the rates range from 3–33% percent, depending on the region.

Malaria is a huge problem for pregnant women in this region of the world. When compared to nonpregnant adults, pregnant women more often have clinical symptoms and are less likely to survive malarial infection.

Schistosomiasis, caused by the genus *Schistosoma*, is a helminthic disease of the cardiovascular system and liver that infects approximately 200 million people in Africa, Latin America, and developing countries elsewhere.

Tuberculosis (TB) is another infection ravaging the developing world. South Africa has the highest rate of TB infection in the world, at more than 300 cases per 100,000 people. Many of these infections are with multiple-drug-resistant strains of *Mycobacterium tuberculosis* (known as MDRTB). (Like HIV, MDRTB is a major disease in the United States, especially in larger cities, such as New York, Miami, Houston, and Philadelphia.)

One of the biggest infectious scourges of the developing world is diarrhea, caused by a variety of microorganisms, notably *Vibrio* species, *Escherichia coli,* and *Salmonella*. It is more deadly for children than for adults, but it is an important source of **morbidity** for all ages.

Answers and Explanations for Case 3.6—In the News

1. What infectious disease do you suppose these cautious citizens were trying to avoid?

Lyme disease. The disease was first reported in 1977, and during the 1990s the number of annual cases doubled. The media carry lots of stories about the disease's possible long-term effects on the joints and nervous system, and people have become understandably cautious about catching it. Most people who acquire the disease have mild forms.

You might have guessed West Nile virus, an infection transmitted by mosquitoes. But the measures taken by people would have been different for a mosquito-borne disease. For instance, they probably would have tried to eliminate all standing water. And there would be no use in erecting fences, which the Massachusetts residents did to keep deer (and their ticks) off their property.

2. What determines which geographical region of the country carries risk of this disease for its inhabitants?

The numbers of ticks in the region. The deer tick (*Ixodes scapularis*) is the usual vector for the Lyme disease bacterium, *Borrelia burgdorferi*. These ticks are most plentiful in the Northeast, but can be seen in the Southeast. *Ixodes pacificus* also can transmit the disease (with more difficulty) and is found in parts of California in fairly significant numbers.

If it's a light year for ticks, even in the mentioned region, it will be a light year for Lyme disease. Ironically, the year this newspaper article was published was a light year for Lyme disease in Massachusetts.

3. Another major disease in the United States is transmitted in a similar way. What is it, and what microorganism causes it?

Rocky Mountain spotted fever, caused by *Rickettsia rickettsii*. This microorganism is transmitted to humans by other kinds of ticks.

4. Which regions of the country have a high incidence of this second tick-borne disease?

Trick question! Despite its name, Rocky Mountain spotted fever is perhaps most common in the Southeast, including North and South Carolina and Tennessee. However, it occurs in every state.

5. Which of these two diseases frequently has no skin manifestations at all?

Lyme disease. Although many people are familiar with the bull's-eye lesion of *Borrelia* infection, in up to half of all cases it never appears.

Answers and Explanations for Case 3.7—Challenge

1. What is an ELISA test, and what does the one for HIV actually detect?

An ELISA (also sometimes abbreviated EIA) is an enzyme-linked immunosorbent assay. It is a "sandwich-style" laboratory procedure in which antibodies must bind to **antigens,** and then be visualized with a colored dye linked to an enzyme. It is used for diagnostic purposes and can be used to detect antigen or antibody. The HIV ELISA detects antibodies to the virus.

If the ELISA is positive it is followed by a more sophisticated test called a Western blot to confirm the positive diagnosis before a patient is notified.

2. This patient did indeed have HIV infection, but continued to test negative. What are some possible explanations for the consistently negative test results?

The tests may have been performed incorrectly, but this is unlikely to happen four times in succession and at two different locations.

HIV test results can also be negative if administered early in the infection, before antibodies to the virus reach detectable titers. This is why repeat testing is recommended. But Fred was tested four times over the course of 9–10 months. **Seroconversion** would certainly have taken place in that amount of time.

It is possible that Fred was infected with a strain of the virus that is not detected (i.e., antibodies to the strain are not screened for) by the HIV kits used in this country. This is a rare occurrence in the United States. Current HIV testing kits are designed to detect antibody to HIV-1 and HIV-2. Other strains have been reported in other countries (notably O strains of HIV). Perhaps he acquired an atypical HIV infection.

Finally, there have been several documented cases of HIV-positive people who simply test negative for HIV antibody. It is not clear why this occurs.

3. Are any alternative tests available to clinicians for patients strongly suspected to be HIV-positive who test negative with the usual test?

In cases where antibody is undetectable, the best strategy is to look for HIV antigen. This removes the "middleman" (the patient's immune response) from the testing process. Blood supplies in this country are tested for an HIV antigen called p24. This test, though not used routinely for diagnosis, can be used in special cases such as Fred's. Other approaches are also available, such as polymerase chain reaction (PCR) testing for HIV RNA or antigens.

4. Would you expect patients with lack of serum reactivity to have a fast or slow progression from HIV infection to AIDS? Defend your answer.

The rare patients who fail to produce antibodies to HIV usually have a very fast progression to severe disease. This might be expected, as there is obviously a defect in the patient's ability to respond immunologically.

5. Which of the reported symptoms are consistent with a diagnosis of HIV?

Extreme fatigue and weight loss are possible signs of HIV infection. Infections that are much more severe or last much longer than usual should also alert clinicians to this possibility. Fred's genital herpes and respiratory infection fit this description. The low CD4 count (normal levels are 1000–1200 cells/ml) and *Pneumocystis carinii* pneumonia are strongly associated with late-stage HIV infection.

Answers and Explanations for Case 3.8—Challenge

1. What do you suppose is happening with Janet? Is it dangerous?

Well, it could be malaria. After all, a mosquito bite led to chills and fever. But the geography is wrong—*Plasmodium* has not been reported to be in the mosquitoes in Michigan. Also, the symptoms around the bite site are not typical of malaria.

It looks like Janet is experiencing **septicemia.** Septicemia refers to the growth of bacteria in the bloodstream. If not treated promptly it can be fatal. Shock can result from the presence of bacteria throughout the cardiovascular system. It can lead to a drop in blood pressure and ultimately death.

2. Explain Janet's symptoms described in the last paragraph of the case.

Janet is sweating yet chilled—indicating she had experienced a fever. Her pale, bluish skin suggests low blood pressure and possible shock. The colored blotches on her skin are often seen with septicemia and indicate localized areas of hemorrhaging caused by the bacteria in the bloodstream.

3. What organism causes this condition?

Many different organisms can cause septicemia. All that is needed is a route into the bloodstream. This can be provided by a simple break in the skin caused by an insect bite, a shaving wound, surgery, and so on. Otherwise-harmless insect bites can sometimes deposit bacteria into the deeper tissues, which results in an infection.

Septicemia can often be seen accompanying infections with *Neisseria meningitidis, Haemophilus influenzae,* and *Pseudomonas* species. *Staphylococcus* and *Streptococcus* are particularly adept at growing in the bloodstream, so they are often found in septicemia as well.

4. When you relate the history of Janet's condition to one of the paramedics, you notice that she writes "secondary to cellulitis" on her pad of paper. What is cellulitis, and what does it mean that Janet's condition is "secondary" to it?

Cellulitis is an infection of the connective tissue underlying skin. The state of Janet's ankle on the day you first went to the hospital indicated cellulitis. Apparently bacteria were introduced by the mosquito bite. Alternatively, the old lotion may have had bacteria in it which Janet then rubbed into the wound. Or perhaps swimming in the lake introduced bacteria into the inflamed wound site.

Without treatment, the bacteria causing cellulitis can rather easily spread to the bloodstream and cause septicemia.

The septicemia occurred after the cellulitis and was related to it. Therefore it was secondary to it.

5. How should Janet's condition be treated at this point?

She should receive intravenous antibiotics. Also recommended are drugs to combat the decrease in blood pressure with the goal of preventing shock. Blood cultures should be performed to determine what the causative organism is and which antibiotic would work best.

Answers and Explanations for Case 4.1

1. What laboratory tests are called for?

Michael's throat should be swabbed twice. One swab should be used to streak a blood agar plate to look for **beta-hemolytic** colonies, which would indicate an infection with *Streptococcus pyogenes*, a group A streptococcus (GAS). The other swab should be used in a 10-minute strep test—an **agglutination** reaction that detects the presence of GAS **antigen** (shown by a color change in the sample compartment). The usual practice is to make the decision to prescribe antibiotics based on the agglutination reaction. The culture results are considered more accurate, but will not be available for 18–24 hours. If the culture results indicate GAS when the rapid test was negative, the physician's office will telephone the patient to offer a prescription.

2. What types of infections are in the **differential diagnosis**?

Sore throats in this age group are usually caused by *Streptococcus pyogenes* or by one of a variety of viruses. The differential diagnosis need only consider whether the infection is streptococcal or not, as strep throat must be treated with antibiotics, whereas all viral throat infections are treated the same—that is, not at all.

3. Your practice has recently been overrun by sore throats and now, late in the evening, there are no supplies for performing the proper test. Should the physician prescribe antibiotics or not?

Infections by group A streptococci can have serious sequelae. They should be treated promptly. In this situation the physician uses the clinical signs and symptoms to decide if the chances are good that it is a streptococcal infection. These include high fever, very bad breath, intense throat pain, and the absence of other upper respiratory symptoms, such as a runny nose.

4. In deciding whether to prescribe antibiotics, should the physician be extra careful not to prescribe an unnecessary antibiotic, or be extra careful not to let a bacterial infection go untreated?

In this case, because the consequences of untreated *Streptococcus pyogenes* infection can be very serious, it is better to **err** on the side of treating with antibiotics. Patients should be counseled to take the drug correctly, and to finish it completely.

5. What are the possible **sequelae** of untreated sore throats?

If the infection is caused by *Streptococcus pyogenes*, complications could include rheumatic fever (with possible heart valve damage), scarlet fever, and, rarely, dangerous bloodstream infections. Viral throat infections usually have no sequelae.

Answers and Explanations for Case 4.2

1. What should your answer to Heather's question be?

No. The vaccine for influenza consists of killed influenza **virions;** it will not cause even a mild case of the flu in any of those receiving the vaccine. Some people report coldlike symptoms after being vaccinated, but remember, the vaccine is usually administered in late fall at a time when respiratory viruses are very common. The fact that a coldlike disease occurs after vaccination is probably due to chance. For a brief period immediately following vaccination, your immune system is in an activated state in response to the antigenic challenge of influenza. In theory this could lead to a slight increase in susceptibility to other microorganisms, but this has not actually been shown to happen.

2. Heather says that because she had a flu shot last year she's going to skip it this year. Respond, with an explanation.

The best protection comes from getting a new flu vaccination every year. The influenza vaccine becomes less effective over the course of time, as a result of changes in the virus. Vaccines cause your specific immune system to respond to surface antigens of microorganisms entering your body. The surface antigens of the influenza viruses in nature are constantly changing in small ways with regard to their chemical composition. This change is called "antigenic drift." So the vaccine you got last year will only alert your specific immune system to recognize last year's surface antigens.

The effectiveness of your immune response depends partly on how good the "fit" is between the antigen and the cells with antigen receptors that your body made after being vaccinated. If the antigen looks a little (or a lot) different this year, it won't fit as well in the preprogrammed antigen receptors and the immune response will be stimulated a little (or a lot) less effectively.

3. What is the difference between "**antigenic** drift" and "**antigenic** shift"?

In contrast to antigenic drift (described in answer 2), antigenic shift is a more abrupt change in antigen composition. In this situation, one or both of the virus's surface antigens (the H spike and/or the N spike) is completely replaced by an H or N spike from an influenza virus that normally infects a nonhuman animal. In that case, there is little hope that our immune system will even partially recognize the totally new spike.

4. What is different about the vaccine from year to year? Who decides what form it will take every year?

The actual strains of virus that are included in the vaccine dose may be different from year to year. The Centers for Disease Control and Prevention (CDC) maintains a worldwide surveillance system that monitors which types of influenza viruses (carrying which type of surface antigens) are causing infections. They decide in the spring of every year which viruses will be most likely to cause fall and winter infections in the United States. Then vaccine manufacturers spend the summer and early fall making enough doses of the vaccine.

5. Susan wants to know why you don't have to get other vaccines annually.

Not all infectious microorganisms change their surface antigens the way influenza does. Changes in surface antigens originate in changes (mistakes) in the genetic material of an organism. Bacteria are generally less likely to make mistakes than viruses, though there are exceptions. And RNA viruses are more likely than DNA viruses to change, because RNA **genomes** are more prone to mistakes in replication than are DNA genomes. Influenza is an RNA virus and its antigenic composition "drifts" naturally over time.

6. Another friend, Dru, says that even though she had the flu shot last year she got terribly sick with the stomach flu over Thanksgiving break and missed most of her vacation. What is your explanation for this?

A lot of conditions are called "the flu," but influenza virus only causes symptoms in the respiratory tract. Intestinal symptoms are not a feature of influenza.

Some mild digestive upset can result from any viral infection, partly because of the action of **interferon** released from your own cells upon viral infection. But a disease that primarily produces digestive tract symptoms is not caused by the influenza virus.

Answers and Explanations for Case 4.3

1. As the physician's assistant in the office, you are the first to examine Doug. What's your tentative diagnosis, based on the history?

Histoplasmosis.

2. Which components of the history support your tentative diagnosis?

Doug is a park ranger; this means he probably spends a lot of time outdoors and may have contact with a variety of animals, including bats and birds, and their droppings, the source of *Histoplasma*. Also, the fact that he has just moved to the area should remind you that a person with no exposure (from another geographical area) may be affected more severely than would a native. Lastly, his infection seems unresponsive to the antibacterial drug.

It has been shown that large percentages of people who live in areas where the fungus is **endemic** have serum antibodies to it, most without remembering any symptoms.

3. Doug's condition has not responded to the antibiotic. List two possible reasons for this finding.

He reports having taken his drug properly, so the first two reasons to consider are (1) that his infection is caused by an antibiotic-resistant strain, and (2) that his infection is caused by a microorganism not susceptible to amoxicillin. Fungal infection should be considered at this point.

4. What are some other conditions caused by this microorganism?

Histoplasma capsulatum can cause severe disseminated infection, but usually only in **immunocompromised** patients. **Chronic** lung infections are also found in this population. It can also cause localized skin or eye lesions. Most people experience asymptomatic infections, after which they have partial immunity to reinfection.

5. Should Julie worry that Doug can transmit the infection to her or to their daughter?

No. Person-to-person transmission of *Histoplasma* has not been documented.

6. What precautions can be taken by other workers who may be regularly or heavily exposed to bat or bird droppings?

Workers can wear respirators when working in areas suspected to have high levels of bat or bird droppings. Additionally, they should wear protective clothing and shoe covers to avoid carrying the infection to their home, car, and other places.

Answers and Explanations for Case 4.4

1. What is the diagnosis?

Hantavirus pulmonary syndrome (HPS).

2. What connection does the diagnosis have with the cabin?

Hantavirus is transmitted through **aerosolization** of the feces, dried urine, or even hairs of infected rodents. Mice (and their products) are very often found in abandoned buildings.

3. You overhear the charge nurse say to herself, "I knew there was a good reason not to clean my house." To what could she have been referring?

Cleaning a building that has not been disturbed for a long time (except by rodents!) can stir up virus-containing materials. If the man or his girlfriend (who admitted she cleans everything) set about cleaning the long-unused cabin, they very likely disturbed quite a bit of dried mouse junk. Without knowing the potential danger of the dust and aerosols, the unprotected couple could have inhaled large quantities of virus.

    The Centers for Disease Control and Prevention (CDC) has special recommendations for cleaning up suspected mouse droppings, or areas where mice have nested. It recommends wetting the entire area down with a detergent or bleach solution, and after a few minutes wiping it up with a damp cloth. Gloves should be worn. Keeping the area wet helps prevent the aerosolization of lightweight, dry mouse materials. When you first enter a building that has been closed up for a long time, the CDC recommends opening windows and doors to allow air to circulate.

4. This case is from Arizona. These infections were first seen in the United States in May of 1993 in the Four Corners area of the Southwest, which includes Arizona, Colorado, New Mexico, and Utah. Can we assume that this disease is only found in the Southwest? What factors determine the places this virus might be present?

Cases of HPS have occurred all over the United States, although a higher density of cases has been reported in the western half of the country. The virus distribution is dependent on the distribution of its animal vector. So far in the United States, four rodents have been found to carry the virus: the deer mouse, rice rat, cotton rat, and white-footed mouse. The geographical distribution of these four rodents covers the entire country. Therefore hantavirus infection may occur anywhere. By the way, the common house mouse seems to be free of hantavirus, but any of the other four rodents could make themselves at home in your home!

Answers and Explanations for Case 4.5

1. What causes mono, or infectious mononucleosis? What do you know about this agent?

Almost always the causative agent is the Epstein-Barr (EB) virus. EB is a member of the herpes family of viruses and shares some of the characteristics of this group, such as **latency.** Once infected with EB, a person retains virus in some cells of the immune system. However, **reactivation** is rare.

2. What are the symptoms?

The symptoms are fever, sore throat, and swollen lymph glands. In addition, many patients experience extreme tiredness.

3. How long will Josh be out of school? Is it okay to visit him?

Josh may be ready to come back to school soon. Most symptoms resolve within a month. After the first two weeks he may feel well enough to attend classes. Returning to basketball practice may take a little longer.

It is safe to visit Josh. The virus is transmitted through saliva, but not through small aerosolized particles that are generated during speaking or breathing. That's why it's often called the "kissing disease."

4. You tease the boys by saying, "Besides, by the time you're adults, all of you will have it anyway." Before they recover from that shock you add, "and some of you have it right now!" Are you just playing around with them, or are these statements true? Explain your answers.

You're telling the truth! In this country up to 95% of middle-aged adults have been infected with EB. Many are infected as young children, and others are infected at a later age, usually with no symptoms. For some reason, nearly half of the people who first acquire the infection during their teenage years experience the **acute** form of the disease, with the symptoms described above. So although it seems as if teenagers are the most frequently infected with EB, they are simply the most visibly infected.

Your comment about some of them being infected now is also true. Because the virus remains in the body in the latent stage, those infected as children will carry the virus. But that's a good thing, as they will have good immunity to reinfection.

5. Sam, the point guard on the team, says his aunt has **chronic** fatigue syndrome. "Isn't that caused by the same virus?" he asks. Is it?

The cause(s) of chronic fatigue syndrome are still elusive. After many years of research, scientists can't pin down one infectious (or noninfectious) cause. Many cases are associated with the presence of EB, but many are not. Occasionally, infectious mononucleosis symptoms do not clear up within two to three months. The condition is then called chronic EBV infection, which resembles, but is distinct from, chronic fatigue syndrome.

Answers and Explanations for Case 4.6

1. What kind of pneumonia is it?

You tell your mother that you didn't catch what your doctor said about the type of pneumonia, either. But you have your book bag with you and you dig out your microbiology text and look up respiratory infections. "RSV!" you say, adding "The symptoms and the treatment match." RSV stands for respiratory syncytial virus. It is the most common cause of respiratory illness in babies under one year of age. It very often results in **bronchiolitis** and pneumonia.

2. Why aren't they giving him antibacterial drugs?

Antibiotics (antibacterials) are **contraindicated** for viral infections of all types. A drug called ribavirin can be delivered via aerosol for **acute** viral respiratory infections.

3. How can the doctor be sure what's causing the pneumonia if she doesn't yet have test results?

She can't be sure, but she can make an educated guess. The main consideration here is the **epidemiology** of the infection. RSV tends to occur in community outbreaks, especially during the cold months, and indeed, the hospital has admitted several babies with the same symptoms in the last 10 days; all of their nasal swab tests were positive for RSV.

There is a rapid test for RSV, but many smaller clinics and hospitals don't yet use it.

4. What about your other child, who is 3 years old? Has she been exposed to the infection by being around the baby? Should the baby remain isolated when he comes home? Can the 3-year-old be vaccinated?

It is very likely that the older sister has already been exposed; it is also likely that she already has antibodies to RSV because most children have had an RSV infection by this age. Many RSV infections in older babies and children are indistinguishable from common cold symptoms.

It is also of little use to keep the two children separate once symptoms have developed in one of them, as transmission is most likely in the four- to five-day incubation period (before symptoms begin).

There is currently no vaccine for this infection, but developing one is a high priority for scientists.

Answers and Explanations for Case 4.7—In the News

1. Health departments often have even less information than this when they have to start hypothesizing about the causative organism and its source. What is your first guess?

Suspect *Legionella* as the causative bacterium. Because the case stated that cleaning the water systems eliminated transmission of the disease, it must have been waterborne. The first organism to suspect in an outbreak of waterborne respiratory disease is *Legionella*.

2. Describe the transmission characteristics of the suspected bacterium.

Legionella species are transmitted in **aerosolized** water, such as that coming from air-conditioning systems, decorative fountains, or water streams used for industrial purposes. Any piece of equipment that produces fine mists of water may harbor and transmit the bacterium.

3. Is there a risk for a continuing community outbreak from these initial infections? Why or why not?

No. *Legionella* is not transmitted from person to person. Infections are acquired via common exposure to an environmental source. Only those exposed to the plant's water systems need worry.

4. Would the health department be likely to identify this bacterium by performing routine water-screening procedures, such as serial dilution or filter collection followed by incubating on EMB or nutrient agar? Explain.

The methods described in this question are appropriate for identifying fecal contaminants, not respiratory pathogens, in water supplies. *Legionella* species do not grow well on common laboratory media. This was one of the reasons it took months to isolate it when it was initially identified in 1976 after it sickened dozens of men attending an American Legion convention in Philadelphia.

The appropriate medium is buffered charcoal yeast agar, which contains critical nutrients, such as iron, needed by the bacterium for growth.

Answers and Explanations for Case 4.8—Challenge

1. Why does the reaction take 36–48 hours to show up?

Because the skin test (also called the Mantoux test) relies on the principle of **delayed hypersensitivity.** T cells that have been sensitized to the antigens during a previous exposure start a cascade of events. It takes one and a half to two days to see the end result of this cascade, which is the infiltration of the area with fluid and defensive cells.

2. If you have a tuberculosis infection, why doesn't the whole body, or at least the respiratory tract, react when the **antigen** is injected during this diagnostic test?

Because the antigens are injected superficially and do not get into the bloodstream. Only a local response, mediated by T cells just under the epidermis, is possible.

3. You are referred for a chest X-ray, but the results are inconclusive. The clinic doctor prescribes a six-month course of isoniazid (abbreviated INH). You take the medicine according to the pharmacist's instruction. Six months later you are taking a medical microbiology course as part of your nursing curriculum. On the day you study tuberculosis, you suddenly realize why you had a positive skin test. It has nothing to do with a true infection, but with the fact that you were born in the Netherlands. Your family moved to the United States when you were 4 years old. What do you suppose is going on here? Discuss as fully as you can.

You received a TB immunization, called BCG, when you were a toddler. Many Western European countries follow this practice, as do other countries around the world. Your T cells had been sensitized to TB antigens through the vaccine.

In the United States, the BCG vaccine is not regularly used. The scenario described here suggests one important reason for this: U.S. public health officials feel it is more important to be able to use the skin test as a reliable screening technique to detect infected individuals. People successfully immunized with BCG would test positive whether or not they are infected with natural *Mycobacterium,* rendering the skin test useless. It is important to note that the BCG immunization procedure has relatively low long-term effectiveness. This fact combined with the loss of screening abilities has led the United States to reject the widespread use of a vaccine.

This case highlights the need to ask patients appropriate questions about their medical histories. All people submitting to TB tests should be asked whether they have ever received the BCG. Because many people don't remember their childhood vaccines, that question should be followed by asking whether they have lived in a foreign country.

4. You have a friend in your hometown who is HIV-positive. When you told her about your TB scare, she said that her specialist can't use the TB skin test, even though HIV-positive people are at higher risk than the healthy population for TB. Why is the skin test not recommended for HIV-positive people?

Because it would be unreliable. Skin test reactivity relies on T cells, which decrease in number during the course of HIV infection. If an HIV-positive person has a negative TB skin test, it may only mean that her T-cell count is very low. In that situation it should not be used to determine whether T-cells are sensitized to TB.

Answers and Explanations for Case 5.1

1. Why steaks instead of hamburgers?

E. coli O157:H7 is a bacterium that contaminates all kinds of food products when they come in contact with the contents of cows' intestines. Meat is contaminated during the slaughter process. Because any given lot of hamburger may be made up of meat from several cows, there is a greater chance that it will be contaminated. A second good reason for ordering steak instead of hamburgers when *E. coli* O157: H7 is a worry is that the bacteria only contaminate the outer surface of animal meats during slaughter. Even the briefest of grilling or broiling will kill surface bacteria on a steak. In ground beef, the "surfaces" of many meat pieces are mixed throughout, so bacteria may be present deep within a hamburger. Therefore, only well-done burgers can be considered safe.

And no, you're not buying. Your good advice is valuable enough!

2. One of your friends acts disgusted and says she'll order a salad instead. Will this guarantee her safety? Why or why not?

No. Any fruit or vegetable grown on a farm that raises cattle or uses cow manure as a fertilizer could have come in contact with the bacterium. Any such product that is not washed thoroughly or cooked before being consumed could potentially transmit this **virulent** *E. coli* strain.

3. One of your friends says that her sister gives her baby apple juice every day. Should she stop? Explain your answer.

There is no threat from any juice or farm product that is pasteurized. Apple cider is different than apple juice. Cider has been a source of *E. coli* O157:H7 infection in the past when it was consumed "raw," or unpasteurized. Some people also enjoy cheeses and dairy products that have not been pasteurized. This practice is now discouraged because of the threat of bacterial contamination.

4. What are the symptoms of *E. coli* O157:H7 infection?

This bacterium is invasive, meaning it can damage the tissues it infects and travel to the bloodstream. This can lead to bloody diarrhea and also bloody urine, because the bacterium attacks red blood cells, which are cycled through the urinary tract and kidneys. Deaths from this infection are often caused by kidney failure. Long-term effects include neural damage and blindness. As with many infections, some people can recover without ever having these symptoms.

5. Another friend says that his family has always eaten rare hamburgers and no one has ever gotten sick. He thinks it's all a bunch of overblown media coverage and says he will continue to eat his favorite delicacy, raw hamburger meat on crackers. What should you tell him?

You should point out that this is not a case of media hype. This particular form of *E. coli* is "new"—it is considered an emerging infection by the Centers for Disease Control and Prevention (CDC). The first cases were recognized in 1982. It seems that this virulent form of *E. coli* has evolved from other *E. coli* that are normal flora in the cow's intestines. It does present a new threat to the public health. Tell him to put (pasteurized) cheese on his crackers instead.

Answers and Explanations for Case 5.2

1. Your Internet search of all kinds of different spellings of "see dif" yields nothing. What section of your microbiology text would likely contain the help you need? What clues lead you in that direction?

The loss of weight, frequent trips to the bathroom, and history of antibiotic therapy should suggest to you that the nurse may be suffering from an intestinal condition. Try the section on gastrointestinal tract or digestive tract diseases. (Of course, you have the extra hint that this case appears in the digestive tract disease chapter!)

2. Now that you've found the right category of infections, can you identify what "see dif" is?

Clostridium difficile (C. diff).

3. Your book has only a small paragraph on this infection. But now you know what to search for on the Internet to find more information. Your instructor wants you to report on the **epidemiology** of the infection. You find that it is referred to as an opportunist and this accounts for its epidemiological patterns. First of all, what is an opportunist?

An opportunist is an organism that may be normal flora (also called normal biota) but that under the correct circumstances can cause disease.

4. Part of an epidemiological description of an infection involves knowing who is most often affected by it. Let's consider opportunistic infections as a group. People in which age groups are most likely to suffer symptoms from an opportunistic infection?

Because the immune systems of young children and the elderly may not be functioning optimally, they are more susceptible to symptoms from otherwise harmless microorganisms. **Opportunistic** infections should always be kept in mind in the **differential diagnosis** for these age groups.

5. In this case the affected nurse is in his mid-30s. Is it his age or something else that predisposes him to the infection? Discuss.

The patient is not in a high-risk age group for opportunistic infections. In this case, the "correct circumstances" (see answer 3) can refer to the fact that the natural defense offered by **microbial antagonism** (also known as "bacterial interference") is disturbed by long-term antibiotic therapy. Not all antibiotic-associated diarrhea is caused by *C. difficile.* But cases that last well beyond the antibiotic therapy could be caused by *C. difficile.*

Another circumstance that can predispose people to opportunistic infections is the suppression of their immune system by infection or by medical intervention (such as treatment with steroids or chemotherapy). This group of patients at high risk for opportunistic infections is often referred to as **immunocompromised.**

6. What is the major **virulence** factor for this microorganism?

C. difficile produces damage to intestinal cells by excreting an **exotoxin.** Finding the toxin in feces is the basis of diagnostic tests for this infection.

Answers and Explanations for Case 5.3

1. First of all, is Pam's husband correct about transmission of the virus? Elaborate.

He is partly correct. The virus is transmitted through contact with bodily fluids of all types including sexual fluids and blood, as mentioned by Pam's husband, but also through saliva and **perinatal** exposure. So, in addition to sexual contact and the use of dirty needles, intimate contact with an infected person or a carrier and being born to a mother with the infection can also transmit the disease. Young children can become infected through normal close contact with an infected caretaker, such as a day-care teacher.

2. How severe is this infection for young children?

Even though the chances for infection at very young ages are relatively low, the outcome of the infection is likely to be most serious in this age group. For instance, infected infants have a 90% chance of developing **chronic** infection, whereas that rate is much lower among those who acquire it as adolescents and adults. The rate of death from liver failure later in life is about 25% among those infected as babies.

Also, because many HBV infections are acquired during adolescence, health officials believe that the best prevention strategy is to administer the full series of vaccinations during well-child visits to the pediatrician during early childhood.

3. Pam says she'll also remind him that in the last year the newspapers have reported at least three hepatitis outbreaks traced back to restaurants. Respond to her statement.

This is not a useful (or accurate!) argument. Restaurant outbreaks of hepatitis are caused by the hepatitis A virus—a completely different virus that is almost never responsible for long-term infections. The hepatitis B vaccine confers no protection from hepatitis A virus.

4. While you're on the phone with Pam, her husband comes home from work. He hears your conversation, and says in a loud voice, "That vaccine is not safe! It's one of those genetically engineered things!" What can you tell Pam about how the vaccine is made, and whether it is safe or not?

The vaccine is considered safe precisely because it is genetically engineered. For one thing, no blood products are used in its manufacture. And the virus is not involved in the production of the vaccine. The gene for the surface **antigen** of HBV is inserted into a harmless yeast cell, which produces large quantities of the surface protein. The protein is separated from the yeast culture and used in the vaccine.

Note: There are rare reports of side effects (the most common is soreness at the injection site) with the vaccine, as with others. But there is no chance of inducing hepatitis B through vaccination.

Answers and Explanations for Case 5.4—In the News

1. Microscopic analysis of the stool samples revealed the presence of small oval-shaped structures, with defined outer walls and two to four nuclei inside that looked like seeds. What is your diagnosis?

Giardiasis, caused by *Giardia lamblia*. This protozoan is often found in its cyst form in stool samples. The cyst has multiple nuclei and is the structure described in question 1.

2. What organisms should be included in the **differential diagnosis** of this infection?

Because many people seem to have simultaneously become sick, it is probable that the outbreak had a common source. (This type of outbreak is called a **point-source outbreak. Epidemics** or outbreaks that are spread sequentially from person to person are termed **propagated outbreaks.**) Some intestinal pathogens that are commonly associated with a single contaminated source, such as food or water, are *Staphylococcus aureus, Salmonella, Shigella, Campylobacter, Escherichia coli* O157:H7, and *Giardia.*

3. What feature of the symptoms suggests that the causative organism is not likely to be *Staphylococcus aureus?*

It is probably not *S. aureus* because gastrointestinal symptoms of that infection, which are induced by a bacterial **exotoxin,** generally last no longer than 24–48 hours.

4. Epidemiologists interviewed the patients about their vacation activities and food intake to try to identify the environmental source of the infection. There was no relationship between illness and a person's attending one of the scheduled children's activities at the hotel. Only two types of food available in the dining room seemed to be associated with the illness: raw vegetables and salads. There was also a statistically significant relationship between illness and having consumed orange juice made from a mix (with hotel water). So what was the likely source?

The water from the hotel is the likely culprit. Food items in the kitchen were washed in it, but were usually cooked afterwards. Only the items consumed raw still posed an infection risk. Likewise, the orange juice was made with contaminated water.

5. Why would an epidemiologist even ask about a person's attendance at children's activities?

Child-care centers and other places where children congregate have been associated with small outbreaks of *Giardia.* This is because children are less **fastidious** about personal hygiene. Objects handled by children with insufficiently washed hands can transmit *Giardia* (and other microorganisms). And it's not just the children! Outbreaks have also been traced back to lunchroom workers who contaminate food with their insufficiently washed hands. Nevertheless, public health officials look for exposures that may have involved children when there is an outbreak of gastrointestinal disease.

6. Are there any symptoms that would help to distinguish this kind of diarrheal illness from others?

Giardia infections can be **acute** (one to three weeks) or **chronic** (months). A large percentage of those infected experience moderate to severe flatulence (intestinal gas). The stool is unusually foul-smelling and can take on a greasy appearance.

Answers and Explanations for Case 5.5—In the News

1. What is the most likely causative microorganism in this outbreak?

Listeria monocytogenes, a short gram-positive rod.

2. Why is this infection associated with processed meats, but usually not with hamburger or cuts of meat including pork, beef, or chicken?

Listeria is a common contaminant of fresh and processed foods, but it is killed by proper heating. Unlike most human pathogens, it grows well at refrigerator temperatures. So foods that are not normally cooked before eating, or those that are often inadequately cooked, may still harbor live bacteria when they are ingested.

Many *Listeria* infections come from the consumption of inadequately cooked hot dogs, especially when they are prepared in a microwave. Microwaving can be a perfectly safe cooking method when done properly. However, inadequate heating times can result in uneven cooking, leaving cool pockets (and living bacteria) within the dense meat material.

3. Epidemiologists describe a microorganism's **pathogenicity** as the proportion of people who become ill after being exposed to the microorganism. (An infection that is **subclinical** in most people who acquire it is considered to have low pathogenicity.) After considering the types of people at high risk for the disease, would you suppose that this organism has high or low pathogenicity? Explain your answer.

Listeria is considered to have a relatively low pathogenicity because most healthy adults with fully functioning immune systems are not made ill by the bacterium even though they are probably exposed to it fairly often. This is evident from the list of people who are at high risk.

Listeria is considered to have a high **virulence,** however. In **epidemiological** terms, virulence is described as the likelihood of serious or life-threatening symptoms once a person has become infected. *Listeria* can lead to the death of infected fetuses or adults, as seen in this case.

We often use the terms *pathogenicity* and *virulence* interchangeably, but you should be aware that they describe distinct aspects of a microorganism.

Answers and Explanations for Case 5.6

1. When the doctor opens the door you whisper that you think there's a case of _____ on the phone.

Cholera

2. The doctor's eyes widen and she asks you how you came to that conclusion. What is your reply?

You explain that the diarrhea is copious, clear, and has mucous flecks but no blood in it. The doctor looks skeptical until you add that the patient is a recent arrival from South America.

3. Why was the doctor initially dubious about your diagnosis and why does the patient's recent immigrant status convince her that your diagnosis was correct?

She is doubtful because cholera is seen relatively infrequently (0–5 cases per year) in North America, especially in a landlocked northern state such as Idaho. Most commonly, cases seen in the United States are associated with recent immigrants or travelers returning from countries where cholera is **epidemic** or still **endemic.** South America has been experiencing an epidemic of cholera since 1991. Idaho has far fewer South American immigrants than do southern states. (Some cases have been associated with eating contaminated fresh shellfish, but usually in coastal areas where large amounts of fresh seafood is consumed.) The fact that the patient lived in Peru until very recently makes cholera a very real possibility, especially with the symptoms you related to the doctor.

4. The doctor asks you to tell Leslie to call 911. The sick woman should be transported to an emergency room right away and the doctor will call ahead and meet her there. What is the first intervention likely to be performed when the patient arrives?

Intravenous rehydration. The symptoms of *Vibrio cholera* are caused by an **exotoxin** that causes a huge influx of fluid into the intestines. The fluid comes from the bloodstream, often leading to severe dehydration, which can result in organ failure and shock and, in 25–50% of the cases, death. The patient must first be rehydrated to prevent the most serious **sequelae** of the disease.

The water loss can also lead to the leg cramps experienced by this woman.

5. The incubation period for this disease is one to four days. Can you think of any way that the young mother could have been infected so recently even though she has been in this country for six weeks?

There are at least two explanations. (1) One of her family members is an **asymptomatic carrier** of the bacterium (a significant proportion of infected people experience no disease). Even though these carriers have no symptoms, they will probably **shed** the bacterium in their feces, providing a possible source of environmental contamination. (2) One of the family members is a **convalescent carrier.** These carriers can shed the bacterium for weeks or months after their own recovery.

6. The next day you ask the doctor about the patient's status. She says that currently the patient is receiving a course of the antibiotic ciprofloxacin, though it won't help her. Why won't it help her and why was it prescribed if it won't?

It is doubtful that antibiotics change the course of this disease. Again, the most effective intervention is rehydration. The drug was prescribed to protect the health of the community and the patient's family, as it will help ensure that shed bacteria are dead bacteria.

Answers and Explanations for Case 5.7

1. After the second stylist walks away, your stylist asks you about hepatitis C. Her first question is, "Is it serious?" Answer this question as thoroughly as you can.

Yes, it is serious. Most people with the infection remain infected their whole life and incur some degree of liver damage over time. On the other hand, some people never suffer ill effects from it, though it is likely that their liver is affected in some way.

Forty percent of **chronic** liver infections are probably due to hepatitis C virus (HCV) infection; chronic liver infections are the 10th leading cause of death in the United States. This translates to approximately 8000–10,000 deaths from HCV per year.

2. How is it transmitted?

This virus is transmitted almost exclusively through blood. Although its transmission and severity resemble hepatitis B, it is *not* transmitted through saliva or sneezing and probably not through sex. The biggest dangers are from injecting **illicit** drugs and from having received tainted blood products in the past. Blood products have been largely free of hepatitis C since 1992.

Keep in mind that there are other ways to come in contact with blood—even through using someone else's toothbrush or razor. Receiving a tattoo or a body piercing from someone not using accepted precautions can also expose you to this and other bloodborne infections.

3. Can she be vaccinated against it?

There is currently no vaccine for hepatitis C. Avoiding contact with blood is the best prevention.

4. Your stylist has heard of hepatitis A and hepatitis B, but never hepatitis C. Is it new? Explain.

The existence of hepatitis C was first established in 1988. Previously, the medical community was aware of a hepatitis not caused by either hepatitis A or B. At that time they named it non-A non-B hepatitis. Before the identification of hepatitis C there was no reliable method for detecting it in blood or tissues, so it was present in some blood products. The infection rate seems to have peaked in the 1980s and has decreased in the 1990s and early 2000s. Approximately 3.9 million Americans are currently infected.

Although the infection rate is declining, deaths from the infection will continue to increase as affected adults reach their 50s, when liver dysfunction often begins to occur.

Answers and Explanations for Case 5.8—Challenge

1. What leads to Guillain-Barré? What would you look for in the patient's history?

Guillain-Barré is usually associated with a previous infectious condition. Most often it is preceded by a viral respiratory condition or a viral or bacterial gastrointestinal disease. It has also occurred after vaccinations. You should look for one of these infections or a history of vaccination in the two- to three-month period preceding onset of Guillain-Barré.

2. Considering this patient's profession, what type of condition do you suspect as the **precipitating** event?

He's a chicken farmer, so you suspect a gastrointestinal infection by *Salmonella* or *Campylobacter.* It is more likely to be *Campylobacter* because as many as one out of 1000 *Campylobacter* infections can lead to some form of Guillain-Barré, and it is possibly the most common infection that can lead to Guillain-Barré in the United States.

3. Is Guillain-Barré often fatal?

No. Most people recover from Guillain-Barré. The condition usually peaks three weeks after the patient first experiences tingling or weakness. In a small percentage of patients, paralysis of the respiratory system can occur. Mortality is significant in this group. Although the most severe symptoms resolve in a few weeks for most patients, residual symptoms may linger for months.

4. What about the original infection, which you identified in question 2? Is it common or uncommon?

Campylobacteriosis is the most common cause of diarrheal illness in the United States. Many cases are **subclinical.** Most people experience typical diarrheal symptoms; bloody stools and fever are a hallmark of this disease.

Here's a good chance to look at the often confusing world of health statistics. Even though *Campylobacter* infections are relatively common, and even though one in 1000 *Campylobacter* infections can lead to Guillain-Barré, the actual incidence of Guillain-Barré is relatively low.

Answers and Explanations for Case 6.1

1. What is the association between belly pain and STDs?

Belly pain is a **sequela** of various bacterial STDs, chief among them gonorrhea and chlamydial infections. It is a possible indication of pelvic inflammatory disease, or PID.

PID is very common. At least one million women per year in the United States have symptomatic PID, and many others may have it without experiencing symptoms.

2. What causes the belly pain, exactly?

The pain comes from inflammation caused by the presence of bacteria in the normally sterile upper reproductive tract, particularly the **endometrium** of the uterus and the **fallopian tubes.** These bacteria, which initially infect the vagina and cervix, may move upward over time or as a result of mechanical displacement, for instance, sexual intercourse or tampon use.

3. Is the belly pain a serious sign?

It can be very serious. PID is the leading cause of female infertility in the United States. In addition, women who have had a single **acute** episode of PID are seven times more likely to experience an **ectopic pregnancy** than women who have never had PID. There are other symptoms, such as difficulty urinating and localized abscesses. Many of these conditions may require surgical treatment.

4. What diagnostic tests are called for? Treatment?

Culture or nonculture tests for *Neisseria gonorrhoeae* and *Chlamydia trachomatis* should be performed. Negative tests for these organisms do not rule out antibiotic therapy for suspected PID, however. The tests may have missed the organisms. Also, other bacteria can cause the condition. If the health care provider suspects PID, appropriate antibiotics should be prescribed even in the absence of information about causative organisms.

If the patient is positive for one of the two organisms above, antibiotics are always indicated.

5. You look further down on the questionnaire and see there is a question about the client's recent sex partners. She answered that she has had relations only with her boyfriend during the past 12 months. Does the boyfriend need information or treatment? Explain.

Sex partners of patients with PID—as well as partners of patients with "uncomplicated" STDs—should also be treated with antibiotics to reduce the possibility of reinfection and to break the cycle of transmission.

Answers and Explanations for Case 6.2

1. How is this infection possible if both of you have been monogamous for at least two years?

You search the Internet for information about gonorrhea and find a toll-free number for STDs that is operated by the Centers for Disease Control and Prevention. There, you speak with someone who explains the following: both gonorrhea (caused by *Neisseria gonorrhoeae*) and chlamydia (caused by *Chlamydia trachomatis*) cause infections of the **urethra** and vagina that may be **acute** or **chronic.** If the infections are acute, resulting in abnormal discharge and pain, infected people are likely to seek medical treatment and, therefore, be diagnosed. But chronic, asymptomatic infections, which are particularly common in women, are not noticeable, and thus may continue for long periods.

In this case, there are two very plausible possibilities: (1) Becky acquired the infection before she met you and because she has not had a pelvic exam in over three years, her infection has gone untreated; (2) you could have had an asymptomatic infection and sometime during the course of your relationship transmitted it to her.

2. What about the immediate problem? Should Becky be treated, even though she is pregnant? Discuss.

Yes. Her obstetrician will know which antibiotics to prescribe. You, too, should receive treatment to avoid reinfecting Becky.

3. Is penicillin the best treatment for gonorrhea? Why or why not?

No. So many *N. gonorrhoeae* isolates are resistant to penicillin and its derivatives that alternative antibiotics should be given. The bacterium may also be resistant to tetracycline, spectinomycin, and even to fluoroquinolones, a class of drugs (including drugs such as ciprofloxacin and ofloxacin) that has been used successfully in recent years. The physician will be familiar with antibiotic resistance patterns in your region and will prescribe an appropriate regimen.

4. Once these facts are explained to Becky, she calms down. But you wonder if there is a shadow of doubt about your fidelity in her mind. You wonder how other couples fare—particularly those who may have less trust in one another, or don't know how to access information about STDs. Speculate about why Becky's physician did not explain all the possibilities.

Talking about sex is difficult for nearly everyone, even physicians. And when interpersonal issues are involved, many doctors feel it is not their place to discuss transmission possibilities. Recent surveys of physicians found that many physicians do not ask their patients about risk factors for human immunodeficiency virus (HIV) and other diseases transmitted sexually. However, this seems to be changing in the face of the HIV epidemic. More physicians are receiving training about the importance of discussing a patient's sexual history and of explaining the more subtle aspects of a patient's positive test results.

Answers and Explanations for Case 6.3

1. What are some of the first questions you should ask?

You should ask:
- (a) Does she have a fever?
- (b) Does she have tenderness in her back (close to her kidneys)?
- (c) Does she have a frequent urge to urinate?

2. The patient reports a fever of 101.5°F and a nearly constant urge to urinate, though she often voids little or no urine. What is your preliminary diagnosis?

Urinary tract infection (UTI), sometimes called cystitis.

3. There is a certain symptom that she has not mentioned. What is it and why is it important that you ask her about it?

Urinary tract infections may progress into infections of the kidney, which are called pyelonephritis. These infections can be suspected if the patient feels midback pain. Physical exams should include percussion (tapping) of the back in the kidney region. Pyelonephritis is a potentially serious development, as inflammatory repair processes in this organ can lead to permanent damage.

4. What is the most likely causative organism for this condition?

Escherichia coli causes nearly 80% of uncomplicated urinary tract infections in healthy women.

5. What is the route of transmission of this organism?

Most commonly, *E. coli,* which are normal flora in the gut, become displaced into the **urethra,** where they cause disease symptoms. This scenario is more common in women than in men, mainly because the anus and the urethra are very close to one another in women, and the length between the **distal** part of the urethra and the bladder is shorter. Men can acquire cystitis through this route (rarely) or from sexual intercourse, either with a woman who is colonized vaginally with these *E. coli* or through rectal intercourse.

6. What are some other causative organisms for this condition?

Proteus, Klebsiella, and *Pseudomonas* are all gram-negative bacteria that can cause UTI.

Answers and Explanations for Case 6.4

1. What is the causative organism of vaginal yeast infections? Where is Jane "getting them from"?

Yeast infections are generally caused by *Candida albicans*. This fungus is normal flora in women during their childbearing years. The numbers of *Candida* are usually kept low by bacterial members of the normal flora, for whom the conditions of the normal vagina are more favorable. When conditions change, *Candida* can overgrow into numbers large enough to cause **acute** symptoms. So it is an **endogenous** or **opportunistic** infection.

Candida is rarely transmitted via sexual contact. However, if a woman experiences active infection her sexual partner may acquire the organism and pass it back to her. Remember, however, that *Candida* is a normal resident of the female vagina. So having it passed to her is not as important in this disease as is her own physiology, levels of normal bacteria, and state of immunity.

2. What conditions could **predispose** a woman to such frequent yeast infections?

Factors that change the hormone concentration or alter the normally acidic pH of the vagina favor overgrowth by *C. albicans*. Often these factors are unidentifiable, but diabetes mellitus, the use of birth control pills, or pregnancy can have this effect and predispose women to frequent yeast infections. The use of broad-spectrum antibiotics (for a different infection) can diminish the normal bacterial population and lead to this condition.

HIV infection can also lead to frequent yeast infections. A healthy T-cell response is required to control *C. albicans*; when T-cell numbers drop during HIV infection, the always-present yeast frequently overgrows the vaginal bacteria, resulting in symptoms. Recurrent yeast infections are often the first sign in women that they may have HIV.

3. Should Jane continue to self-medicate for her yeast infections or should she see a doctor? Please explain.

She should see a doctor, especially if she does not have any of the risk factors for increased susceptibility to yeast infections listed in answer 2. The physician should review her history, examine vaginal **exudate** for the presence of large numbers of yeast, and determine whether an HIV test should be performed.

Even in the absence of HIV infection, it is advisable for Jane to get physician care to ensure that she is treating the infection properly with over-the-counter medications.

4. Are there any possible serious consequences of vaginal yeast infections?

Very rarely, and mainly in **immunocompromised** people, *C. albicans* infections of body surfaces can progress to bloodstream infections, which are very difficult to treat and often fatal.

Answers and Explanations for Case 6.5

1. What usually causes bacterial vaginitis (often called BV)?

Many different anaerobic bacteria can cause this condition. The most common cause is *Gardnerella vaginalis,* though *Mycoplasma* or *Prevotella* species can also cause it. In this condition the normal, aerobic *Lactobacillus* species are replaced with anaerobic bacteria.

2. What are clue cells? What bacterium is associated with clue cells?

Sloughed vaginal epithelial cells are called clue cells when they have small bacteria attached to them in distinctive patterns. The clue cells bear **gram-variable** bacteria, often in pairs or small groups. The bacteria should not resemble gram-positive cocci. They are most often *Gardnerella vaginalis.*

3. Is BV a dangerous condition? Explain.

BV is often asymptomatic in women. It can result in white discharge (from the inflammatory process initiated by the bacteria in the vagina). Although it is usually of no consequence for the women who experience it, it is potentially dangerous to the fetus, as it has been associated with premature delivery and neonatal **septicemia.**

4. Is BV a sexually transmitted disease? Elaborate.

BV is found more often in women who have had multiple sexual partners, but has not been specifically found to be transmitted sexually. Some of the organisms that can lead to BV are found in large percentages of healthy women.

5. What other diseases are in the **differential diagnosis** of a woman with copious vaginal discharge?

Chlamydia and gonorrhea can cause a discharge that is often difficult to distinguish from normal vaginal secretions. *Trichomonas* infection results in frothy greenish discharge, and *Candida albicans* can result in discharge resembling that of BV. As a first step, microscopic examination of the vaginal fluids should be performed. This can rule out *Candida* and *Trichomonas,* as both of these are readily seen on a stained preparation. When clue cells are present as more than 20% of the vaginal cells, BV caused by *Gardnerella* can be presumed.

When no clue cells are present and the other organisms are not visible, physicians often perform a KOH (potassium hydroxide) test. In this test, one drop of 10% KOH is added to a wet preparation of vaginal fluids. If a fishy odor results immediately, *Gardnerella* BV is diagnosed. Further diagnostic tests for the presence of *Chlamydia* and *Neisseria gonorrhoeae* may be performed.

6. Is BV treatable? If so, with what? What is the likely outcome of treatment?

Yes. This condition is usually treated with the antibiotic metronidazole. Pregnant women with BV should receive this treatment in order to greatly reduce the chance of pre-term delivery.

Answers and Explanations for Case 6.6

1. Are they safe if she does not have lesions at the time of their intercourse? Why or why not?

No. Recent studies have found that people infected genitally with the herpes simplex virus can still **shed** the virus (release it from their bodies to the environment) when there are no visible lesions.

That may be one of the reasons the herpes infection rate is so high. Many people (some scientists say the majority of those infected) do not even know they are infected because they never have lesions. But they might be able to transmit the virus anyway.

2. Whether Jack's girlfriend has lesions or not, if he uses a condom he will be protected, right?

Wrong. Condoms only protect the areas that are covered by the latex. For some diseases, in which the female carries the infection in the vagina or **cervix,** this can be an effective way to halt the transmission to a male, as the only part of the male coming in contact with infected tissue is the covered part of the penis. But herpes lesions may be on the external genitalia of the female, which could contact the uncovered genitals of the male.

3. If Jack were the one with herpes and his girlfriend was uninfected, would his use of a condom completely protect her? Explain.

No, but the risk of transmitting the virus from an infected male to a female is somewhat lower if the male has lesions only on the shaft of his penis, which will be covered by the condom. If the male has lesions in areas left bare by a condom, his condom will not protect his female partner. It is recommended that infected people not have sex if there are visible lesions, even with a condom.

4. What would you say is the safest way for Jack and his girlfriend to have intercourse?

"Safe" is a matter of degree; the only truly safe sex is no sex. But no matter who the infected partner is, the use of a female condom may provide better protection for the uninfected partner. In addition to covering her vaginal walls and cervix, it extends to cover parts of the external genitalia.

5. What about those new drugs Jack has heard about on TV? Can his girlfriend take those and cure herself? Or at least avoid infecting him? Give some detail.

Several antiviral medications now on the market are used for genital herpes: valacyclovir (trade name Valtrex), acyclovir (Zovirax), and famciclovir (Famvir). These are never curative. They are likely to decrease the numbers of **virions** that become activated, thereby shortening the duration of lesion episodes, or possibly preventing the lesions altogether. But they do not eliminate all virions and thus do not eliminate the possibility of transmission.

Answers and Explanations for Case 6.7—Challenge

1. What kind of infectious diseases come to mind when a widespread rash is seen as the primary complaint? (Hint: Why had the doctor asked about her sexual history? Why did he ask about her immunizations?)

Chicken pox, measles, scarlet fever, fifth disease, meningococcal meningitis, and secondary syphilis, among others. The patient says she has never had sexual intercourse, so syphilis is ruled out. Because the patient's **MMR** is up to date, measles is considered less likely. Fifth disease is very mild and not likely to cause **acute** symptoms.

2. Her rash was diffuse, with well-separated bumps that were **maculopapular.** Was it likely to be chicken pox? Why or why not?

No. Chicken pox causes clear, fluid-filled, **serous** lesions. The fact that the bumps are well separated makes scarlet fever unlikely as well, as the rash associated with that disease produces smaller red bumps that seem to merge with one another, giving the whole skin a bright red appearance.

3. The cerebrospinal fluid obtained from the lumbar puncture was clear—no evidence of bacteria. Another infection was ruled out. Which one?

Neisseria meningitides (meningococcal) meningitis. This was in the **differential diagnosis** because it is often accompanied by a skin rash.

4. The doctor then asked the patient about her menstrual history and practices. She began menstruating at the age of 12 and reported that her last period began four days ago. She reported that she mainly uses tampons during her period. What infection do you think the doctor had in mind in asking about menstruation? What do you know about the infection in question?

Staphylococcal toxic shock syndrome (TSS). In 1980, the incidence of TSS began to rise drastically, and cases were associated with the use of highly absorbent tampons. The patient reported using these tampons during the last four days. The doctor sent her to the hospital where blood cultures were positive for *Staphylococcus aureus*. After seven days on intravenous antibiotics, the infection was cleared.

5. Your mother says that if you see a patient with these symptoms once you start your practice as a physician's assistant, it is less likely to be the same infection. Why?

The overall incidence of TSS decreased rapidly once tampons were recognized as a **predisposing** factor. The most highly absorbent ones were removed from the market, and public education campaigns alerted women to the danger of not changing tampons frequently.

Nonmenstrual TSS is still seen. It has occurred (rarely) after insertion of intrauterine devices (IUDs) and can occur in both men and women as a result of surgical infection. It should be noted that another coccal bacterium, *Streptococcus pyogenes*, can cause a TSS-like illness. It may follow *S. pyogenes* infection of broken skin or, less frequently, the pharynx.

Answers and Explanations for Case 6.8—Challenge

1. What questions about the patient's behavior should the physician ask during the history?

The patient should be asked about recent sexual activity. It is highly likely that the genital lesions were transmitted through sexual contact. This information helps the doctor make a diagnosis. In addition, the patient can be counseled to inform her partners that they, too, may have the infection.

2. What is your **presumptive diagnosis** based on the facts presented? What other conditions might be in the **differential diagnosis**?

This is most likely a primary occurrence of genital herpes, caused by herpes simplex virus type 1 or type 2 (HSV-1 or HSV-2). Most, but definitely not all, genital lesions are caused by HSV-2. Fever and **acute** illness sometimes accompany first episodes of the lesions. Less frequently, meningitis may be seen, as appears to be the case here. It should be noted that initial infection can also be unnoticed and the infection may continue without outbreaks of lesions. Therefore, a large number of people are infected with HSV without being aware of it.

Other lesion-inducing diseases should be considered in the differential diagnosis. Chancroid, caused by *Haemophilus ducreyi*, is often accompanied by pronounced swelling of the **inguinal** lymph nodes. Genital warts cause lesions, but they are not fluid-filled lesions and are generally not painful. The primary chancre of syphilis should be considered, even though those lesions are usually larger than the 2–3 mm described here.

To make things more complicated, you also need to consider the possibility that the genital lesions and the mild meningitis symptoms seen here may be separate, unrelated diseases.

3. Why do you suppose the patient was treated with an antiprotozoal drug after her first abnormal Pap?

Presumably, the patient was treated with an antiprotozoal because her physician suspected *Trichomonas vaginalis* infection. Pap smears detect abnormal morphology of cervical cells, which may be caused by a variety of conditions. Pap smears were developed to screen for cervical cancer, but other conditions can be signaled by abnormal cells as well. Physicians sometimes prescribe antimicrobial treatment based on pathology reports accompanying Pap smear results.

4. Of what importance is the patient's history of abnormal Pap smears?

The abnormal Pap smear probably is not related to the patient's current condition. As noted above, abnormal Pap smears could signal a variety of conditions, including various infections. The patient said she finished the antibiotic course prescribed for her after her first abnormal Pap; if she had *Trichomonas* it should have been cured. The second abnormal Pap smear indicates some other condition, possibly human papillomavirus (HPV). HPV infection can lead to cervical cancer, the early signs of which are detected with Pap smears. However, HPV does not cause lesions of the type described. Another etiological agent must be considered.

5. What tests should be ordered to confirm the presumptive diagnosis?

Culturing HSV from the lesions is the only way to definitively diagnose HSV infection. However, most clinical laboratories do not do extensive viral culturing and most physicians who want confirmation of their presumptive diagnosis order antibody tests. This requires drawing the patient's blood and looking for antibody to HSV-1 or HSV-2. The lesions are so distinctive and the **epidemiology** of HSV so well established that many physicians do not order any tests.

The Centers for Disease Control and Prevention recommends, however, that whenever genital lesions are present, diagnostic tests for syphilis, chancroid, and HSV be performed. Keep in mind that it is not unusual for a person to be infected with more than one of these STDs.

Glossary

Active immunity the ability of a host to mount a specific immune response whenever exposed to a particular antigen

Acute of short duration, rapid and abbreviated in onset, in reference to a disease

Acute and convalescent sera serum drawn from a patient at two times: when the patient is experiencing symptoms of a disease and at some time afterward, in order to detect changes in levels of serum antibody; also called *paired sera*

Aerosolization the dispersal (as in infectious agents) in the form of a fine mist or spray

Agglutination a reaction in which particles (such as red blood cells or bacteria) suspended in a liquid collect into clumps; occurs especially as a response to a specific antibody

Alpha-hemolytic α-hemolytic; a characteristic of bacteria visible on blood agar plates in which there is incomplete clearing around a colony as a result of the action of bacterial enzymes that diffuse into the agar; clearing appears green or brownish; see also beta-hemolytic and gamma-hemolytic

Anatomical diagnosis the identification of the physical site of symptoms

Antigen a substance, usually protein or carbohydrate (as a toxin or enzyme), capable of stimulating an immune response

Antigenic capable of stimulating an immune response

Asymptomatic carrier a person infected with a microorganism but showing no effects

Axillary relating to or located near the axilla, which is the armpit

Beta-hemolytic β-hemolytic; a characteristic of bacteria visible on blood agar plates in which there is complete clearing around a colony as a result of the action of bacterial enzymes that diffuse into the agar; clearing appears transparent; see also alpha-hemolytic and gamma-hemolytic

Bilateral having, or relating to, two sides

Bronchiolitis inflammation of the membranes lining the bronchioles

CD4 count number of CD4 or T_{helper} cells in a milliliter of blood; used as an indicator of progress of HIV infection

Cervix the narrow lower or outer end of the uterus

Chemotherapy the use of chemical agents in the treatment or control of disease or mental disorder; in common use refers to treatment of malignancies

Chocolate agar a semisolid medium made by adding heated blood to nutrient agar

Chronic of long duration or frequent recurrence, referring to a disease or ailment

Communicable capable of being transmitted from one host to another; contagious

Contraindicate to make (a treatment or procedure) inadvisable

Convalescent carrier a person who has recovered from an infection but is still colonized by the microorganism

Copious large in quantity; abundant

Debilitating describes a condition that saps the strength or energy, severely reducing patient's function

Debride to cleanse by surgical excision of dead, devitalized, or contaminated tissue and removal of foreign matter from a wound

Delayed hypersensitivity hypersensitivity (as in a tuberculin test) that is mediated by T cells; typical symptoms of inflammation and **induration** (hardening) appear after an interval of 12–48 hours following exposure to an antigen (as by injection of the antigen under the skin), provided the individual has previously been exposed to the antigen

Dermatophytes types of fungi parasitic upon the skin or skin derivatives (hair or nails)

Differential diagnosis a list of diseases or conditions presenting similar symptoms

Distal anatomically located far from a point of reference, such as an origin or a point of attachment; often refers to the portion farthest from the center of the body

Dormant in a condition of biological rest or inactivity; the state of being stopped in growth or development

Dysfunction difficult function or abnormal function

Ectopic pregnancy implantation and subsequent development of a fertilized ovum outside the uterus, as in a fallopian tube

Endemic prevalent in or peculiar to a particular locality, region, or people

Endogenous from within

Endometrium the glandular mucous membrane that lines the uterus

Epidemic an outbreak of a contagious disease that spreads rapidly and widely; incidence of disease above normal levels

Epidemiology the branch of medicine that deals with the study of the causes, distribution, and control of disease in populations

Epiglottitis respiratory disease caused by *Haemophilus influenzae;* results in respiratory obstruction that leads to death in nearly all untreated patients

Epithelium membranous tissue composed of one or more layers of cells forming the covering of most internal and external surfaces of the body and its organs

Err to make a mistake or error

Etiological diagnosis the determination of the root cause of a disease; if disease is infectious this involves identifying the causative microorganism

Etiology the cause or origin of disease

Exotoxin a poisonous substance secreted by a microorganism and released into the medium in which it grows

Exudate a substance that has oozed forth

Fallopian tubes the tubes or ducts for the passage of eggs from the ovary to the uterus where further development takes place; sometimes called uterine tubes

Fastidious being particularly careful about cleanliness and hygiene (when used as a description of microorganisms, this term means requiring multiple complex nutrients and/or special conditions to grow)

Fomite an inanimate object or substance that is capable of transmitting infectious organisms from one individual to another

Gamma-hemolytic γ-hemolytic; describes bacteria that do not possess enzymes capable of destroying red blood cells on a blood agar plate, resulting in no clearing around a colony; see also alpha-hemolytic and beta-hemolytic

Genome the complete complement of genetic material in an organism; in bacteria and fungi may include plasmids as well as chromosomes

Gram-variable describes an organism that may stain gram-positive or gram-negative or both

Herd immunity the prevention of the spread of disease by limiting the number of susceptible hosts

Histological pertaining to the anatomical study of the microscopic structure of animal and plant tissue

Household contacts people who share the same house as an infected person

Illicit not sanctioned by custom or law; unlawful

Immunocompromised incapable of developing a normal immune response, usually as a result of disease, malnutrition, or therapy to suppress the immune response

Inguinal   pertaining to the groin

Interferon   any of a group of glyco-
proteins that act to prevent viral
replication; produced by human
cells in response to infection by
a virus

Intermittent   stopping and starting at
intervals

In utero   in the uterus

Keratinized   to possess keratin, a
tough, insoluble protein found on
hair, skin, and nails

Labia   any of four folds of tissue of the
female external genitalia

Latency, latent   in a dormant or hid-
den stage

Localized   confined to a specific area
of the body; not **systemic** or generalized

Lumbar puncture   the insertion of a
hollow needle beneath the arach-
noid membrane of the spinal cord
in the lower back region to with-
draw cerebrospinal fluid or to
administer medication

Lymphocyte   a white blood cell that
functions in specific defenses of the
host

Lysozyme   an enzyme occurring naturally in human tears, saliva, and
other body fluids, capable of
destroying the cell walls of certain
bacteria and thereby acting as a
mild antiseptic

Maculopapular   describes a small,
solid, usually inflammatory elevation
of the skin that does not contain
pus; area of discoloration on the
skin caused by excess or lack of
pigment

Malaise   a vague feeling of bodily
discomfort, often accompanying
illness

Microbial antagonism   the protective
effect provided by normal microbial
biota on human tissues in which
transient pathogenic microorganisms have limited opportunity to
colonize

MMR   immunization against measles,
mumps, and rubella

Morbidity   illness

Mucocutaneous   referring to the skin
and/or mucous membranes

Murmur   an abnormal sound, usually
emanating from the heart, that
sometimes indicates a diseased condition

Mycosis (pl.: mycoses)   any disease
caused by a fungus

Niche   a place, defined by space, time,
or circumstances, in which an
organism prospers

Notifiable disease   certain diseases
that when identified must be
reported to the local and/or national health authorities

Occlude   to close off

Opportunistic   describes microorganisms that are typically not pathogenic but take advantage of a change in host condition, a new location within a host, or transmission to a more susceptible host to cause disease

Oral history   portion of a clinical examination in which the health care provider asks pertinent questions about the patient's symptoms and past and present exposures and conditions

Otitis media   inflammation of the middle ear, often caused by infection

Passive immunity   immunity acquired by the transfer of antibodies from another individual, as through injection or placental transfer to a fetus

Pathogenicity   ability to produce disease; with microorganisms calculated as the proportion of people who become ill after being exposed to the microorganism

Perinatal   the period around childbirth, especially the five months before and one month after birth

Peripheral   the surface or outer part of a body or organ; external; in the case of the nervous system refers to nerves radiating from the spinal cord

Petechial   marked by a small purplish spot on a body surface, such as the skin or a mucous membrane, caused by a minute hemorrhage

Pleomorphic   having two or more structural forms during a life cycle; for example, a bacterial species that exhibits both coccus and bacillus morphology

PMN   polymorphonuclear cell; a white blood cell that functions in nonspecific host defenses

Point-source outbreak   an infectious disease outbreak in which all patients are infected by the same source, such as contaminated water

Portal of entry   the route by which an agent gains access to the body

Precipitating   leading to, as in *precipitating event,* the event that led to the current condition

Predispose   to make susceptible or liable

Present   (in clinical practice) to make oneself present to a health care provider

Presumptive diagnosis   the best guess about the nature or cause of a disease before definitive answers are available

Propagated outbreak   an infectious disease outbreak in which original patient(s) infect additional patients in a continuing fashion

Prophylactic   acting to defend against or prevent something, especially disease; protective, such as a vaccine or drug

Prostration   total exhaustion or weakness; collapse

Reactivation   in infectious conditions, the reemergence of symptoms due to increase in numbers or activity of a latent microorganism

Recombinant   pertaining to a genome that is genetically engineered; an organism that has foreign genetic material incorporated into its genes

Reservoir   any person, place, or thing that is a source of an infectious agent

Septicemia   a systemic disease caused by the multiplication of pathogenic organisms or the presence of their toxins in the bloodstream. Also called blood poisoning

Sequelae (sing.: **sequela**)   pathological conditions resulting from a disease

Seroconversion   development of antibodies in blood serum as a result of infection or immunization

Serous   containing, secreting, or resembling serum

Shed   to give off or emit infectious agents in some way

Sloughed   describes cells separated from surrounding living tissue and released into the environment

Subclinical   not manifesting characteristic clinical symptoms; referring to symptoms that are unnoticeable

Systemic   affecting the entire body or an entire organism

Triage   a process for sorting injured or ill people into groups based on their need for immediate medical treatment

Urethra   the canal through which urine is discharged from the bladder and through which semen is discharged in the male

Virions   complete viral particles, consisting of RNA or DNA surrounded by a protein shell; the infective form of a virus

Virulence   degree of damage that can be produced by an infectious agent

Virulent   capable of inflicting moderate to severe damage to a host

Zoonosis (pl.: **zoonoses**)   a disease of animals that can be transmitted to humans

Index